# SpringerBriefs in Intelligent Systems

Artificial Intelligence, Multiagent Systems,
and Cognitive Robotics

**Series Editors**

Gerhard Weiss, Maastricht University, Maastricht, The Netherlands
Karl Tuyls, University of Liverpool, Liverpool, UK; Google DeepMind,
London, UK

This series covers the entire research and application spectrum of intelligent systems, including artificial intelligence, multiagent systems, and cognitive robotics. Typical texts for publication in the series include, but are not limited to, state-of-the-art reviews, tutorials, summaries, introductions, surveys, and in-depth case and application studies of established or emerging fields and topics in the realm of computational intelligent systems. Essays exploring philosophical and societal issues raised by intelligent systems are also very welcome.

More information about this series at http://www.springer.com/series/11845

M. N. Murty · Anirban Biswas

# Centrality and Diversity in Search

## Roles in A.I., Machine Learning, Social Networks, and Pattern Recognition

M. N. Murty
Department of Computer Science
and Automation
Indian Institute of Science
Bengaluru, Karnataka, India

Anirban Biswas
Department of Computer Science
and Automation
Indian Institute of Science
Bengaluru, Karnataka, India

ISSN 2196-548X                 ISSN 2196-5498   (electronic)
SpringerBriefs in Intelligent Systems
ISBN 978-3-030-24712-6         ISBN 978-3-030-24713-3   (eBook)
https://doi.org/10.1007/978-3-030-24713-3

This Springer imprint is published by the registered company Springer Nature Switzerland AG
The registered company address is: Gewerbestrasse 11, 6330 Cham, Switzerland

# Preface

## Overview

*Centrality and Diversity* are two important notions in Search in a generic manner. Their *Roles in A.I., Machine Learning (ML), Social Networks, and Pattern Recognition* are important. This book aims at clarifying these notions in terms of some of the foundational topics like *search, representation, regression, ranking, clustering, optimization,* and *classification.*

Centrality and diversity have different roles in different tasks associated with *AI* and *ML*. For example, search may be generically viewed as playing an important role in

- *AI* problem solving. Here, we represent a problem configuration as a state and we reach the *goal state or final state* by using appropriate search scheme.
- Representation of a problem configuration in *AI*, representation of a data point, class, or cluster.
- Optimization which itself involves the search for an appropriate solution.
- Selecting a model for classification, clustering, or regression.
- Search engines where the search is the most natural operation.

Representation itself is an important task in a variety of tasks. Popularly representation deals with every task in *AI* and *ML*. Optimization is controlled through some regularizer to reduce the diversity in the solution space.

Clustering is an important data abstraction task that is popular in *ML*, data mining, and pattern recognition. Classification and regression have some common characteristics and *bias–variance trade-off* unifies them. Ranking is important in a variety of tasks including *information retrieval.*

*Centrality and diversity* play different roles in different tasks. In classification and regression, they show up in the form of variance and bias. In clustering, centroids represent clusters and diversity is essential in arriving at a meaningful partition. Diversity is essential in ranking search results, recommendations, and summarization of documents.

## Audience

This book is intended for senior undergraduate and graduate students and researchers working in machine learning, data mining, social networks, and pattern recognition. We present material in this book so that it is accessible to a wide variety of readers with some basic exposure to undergraduate level mathematics. The presentation is intentionally made simpler to make the reader feel comfortable.

## Organization

This book is organized as follows:

Chapter 1 deals with a generic introduction to various concepts including centrality, diversity, and search. Further, their role in several *AI* and *ML* tasks is examined. Chapter 2 deals with searching and representation is discussed in Chap. 3.

Clustering and classification form the subject matter in Chap. 4. Ranking is examined in Chap. 5. Chapter 6 deals with applications to social and information networks. Finally, it is concluded in Chap. 7.

Bengaluru, India                                                      M. N. Murty
                                                                      Anirban Biswas

# Contents

# Acronyms

| | |
|---|---|
| AI | Artificial Intelligence |
| CC | Clustering Coefficient |
| DTC | Decision Tree Classifier |
| JC | Jaccard Coefficient |
| KLD | Kullback–Leibler Divergence |
| KMA | K-Means Algorithm |
| KNNC | K-Nearest Neighbor Classifier |
| LDA | Latent Dirichlet Allocation |
| LP | Link Prediction |
| LSA | Latent Semantic Analysis |
| MDC | Minimal Distance Classifier |
| MI | Mutual Information |
| MLP | Multilayer Perceptron |
| NBC | Naïve Bayes Classifier |
| NNC | Nearest Neighbor Classifier |
| PCA | Principal Component Analysis |
| SVD | Singular Value Decomposition |
| SVM | Support Vector Machine |
| TF-IDF | Term Frequency–Inverse Document Frequency |

# Chapter 1
# Introduction

**Abstract** Search is an important operation carried out by any machine learning task. Centrality and diversity play a potential role in every machine learning task. This chapter introduces the role of centrality and diversity in search carried out by a variety of tasks in machine learning, data mining, pattern recognition, and information retrieval.

**Keywords** Search · Centrality · Diversity

## 1.1 Introduction

Search is easily the most important and popular task carried out by any digital computer. In an extreme view, any operation on a computer involves search. For example:

- Finding the optimal value of a function might amount to *searching for a solution* based on a gradient descent procedure or some other variant. So, *optimization is search* only. Gradient-based searching is popular in *deep learning*.
- In *machine learning*, we infer a model by using a learning algorithm and data. Such an inference involves search. For example, in a linear classifier like the *support vector machine (SVM)*, we find a weight vector $W$ and a threshold $b$ from the training data. Intrinsically, it involves searching for $W$ and $b$ corresponding to maximizing the margin which involves search again. In the process, the support vectors with their associated weights will be searched for and these are used to get $W$ and $b$.
- Matching is an important operation often exploited by machine learning algorithms. It could be either exact or approximate matching based on the need. Further, it could be matching strings, vectors, trees, and the like. Note that every such matching operation is also a search operation.
- In the areas of machine learning and deep learning, search is a popular activity. It could be in the form searching for an optimal *representation* of patterns/data/ classes/clusters; it could be searching for the right set of *parameter values* which

is optimal in some sense; it could be searching for the optimal *matching/proximity function*; or it could be in *model selection and assessment*.

- In information retrieval or for *search engines*, search is the *most fundamental* operation. Here, results obtained based on an appropriate search operation performed over a collection of documents are ranked. So, this search plays a vital role in the *overall ranking* as well.
- In the most extreme case, even *basic numerical operations involve search*. For example, if the sine or cosine function values for several arguments are required, then these function values could be obtained by a *table lookup or search* operation. This could be used to reduce the repeated computational overhead.

## 1.2   Notation and Terminology

First we describe the terms that are important and used in the rest of the book.

- **Pattern**: It is a physical object like a dog, a pen, *etc* or a property of objects like signature, speech, iris, fingerprints of humans *etc*.
  Because we are dealing with machine learning, we often do not directly deal with the objects, rather we deal with their representations. Even though pattern and its representation are different, it is convenient and customary to use *pattern* for both. Further, we sometimes use equivalent terms like point, data point, object, example, and vector also to mean the same. The context clarifies the usage.
- **Feature**: It is a property/characteristic of the pattern. These features are also called *attributes*. These could be either *numerical or categorical* in general. However, in most of the current applications, only numerical features are used.
- Even though there could be other ways of representing patterns, we use the *vector representation* of patterns with a view that the patterns are elements of a vector space.
  There could be some important applications where we encounter categorical features, for example, terms in documents. It is customary to convert them into numeric vectors in areas like *deep learning*.

- **Collection of Patterns**: A collection of $n$ patterns, $\mathcal{X}$, is a set $\{x_1, x_2, \ldots, x_n\}$, where $x_i$ denotes the $i$th pattern.
  We assume that each pattern is an $l$-dimensional vector unless otherwise stated. So, $x_i = (x_{i1}, x_{i2}, \ldots, x_{il})$.

- **Cluster**: It is a collection of *similar* patterns.

  - Similarity or proximity is captured either explicitly through some function that typically maps a pair of vectors to a nonnegative real number or it could be *implicit* where some *gestalt* property underlies the characterization of the cluster as a whole. In a simple sense a cluster is a set of the points in it. This is popular in *hard clustering*.

- *Hard clustering*: Here, clustering is partitioning where a partition of a collection $\mathcal{X}$ is

$$\pi(\mathcal{X}) = \{c_1, c_2, \ldots, c_K\}.$$

Here $c_i \subset \mathcal{X}$ is the $i$th cluster, $\bigcup_{i=1}^{K} c_i = \mathcal{X}$ and $c_i \cap c_j = \phi$.
- *Syntactic Labeling*: This $K$-partition of a collection of $n$ points $\mathcal{X}$ is such that the label $i$ associated with cluster $c_i$ is *syntactic*. So, the labels of clusters $c_i$ and $c_j$ could be swapped without loss of information.
- *Soft clustering*: Here, it is possible that the same point $x_j$ could belong to more than one cluster. So, $c_i \cap c_j = \phi$ is not satisfied.
- *Description of a cluster*: It is useful to describe a cluster by using some kind of summary. In statistical clustering, a cluster could be described by its centroid which is a kind of summary where

$$centroid(c_i) = \frac{1}{|c_i|} \sum_{x \in c_i} x.$$

In logic-based clustering, a cluster is described by a *concept* which is a logical formula. In *topic modeling*, a topic is an assignment of probabilities to the components of the vectors in the cluster. So, the underlying soft cluster is defined by this probability distribution.

- **Class**: A class is a collection of points that are associated with some *semantic label*. So, $i$th class $c_i$ and $j$th class $c_j$ carry different semantics. So, they are associated with different *semantic labels*.

We use some other terms also in the later chapters and their usage will be clear based on the context.

## 1.3 Centrality and Diversity

So search plays a vital role not only in digital computer operations but also in machine learning, data mining, pattern recognition, and information retrieval in the form of important generic operations like *representation, inference, matching, optimization, and parameter tuning*. More specifically, these operations are involved in machine learning tasks including *representation of patterns, clusters, classes; clustering; classification; ranking; and regression or curve fitting*. We will examine how generic notions *centrality and diversity* play a role in each of these tasks. We will see how they play a specific role in each task before abstracting these two concepts.

## 1.3.1  Representation

We represent patterns as vectors and each component corresponds to a feature. Based on the type of application and patterns, the dimensionality could vary. For example, consider the simple $4 \times 4$ binary patterns shown in Fig. 1.1. There are four patterns labeled (a), (b), (c), and (d) in the figure. The figure in (a) may be viewed as the digit 1 and the one in (b) is a version of digit 7. Similarly, the patterns in (c) and (d) may be viewed as noisy variants of digits 1 and 7, respectively. It may be possible to depict the characters better using bigger size patterns. However, for the sake of illustration, we are using these $4 \times 4$ patterns.

It is possible to consider each of the patterns as a 16-dimensional binary pattern by considering the row-major order, where rows follow each other from top to bottom in a sequence. For example figure (d) is viewed as "1 1 1 1 0 0 0 1 0 0 0 1 0 0 1 1". The four patterns may be viewed in 16 dimensions as shown in Table 1.1. Note that the Hamming distance between the 16-bit strings corresponding to (a) and (b) is 3 as they differ in the first three bits. Similarly, strings against (a) and (c) differ in 1 bit, the 13th bit, and so the Hamming distance between (a) and (c) is 1. Hamming distance between various pairs are depicted in Table 1.2. Here each row and each column is labeled by one of the four patterns. The $ij$th entry in the table indicates the Hamming distance between the $i$th and $j$th patterns, where $i$ is the row label and $j$

```
(a)                    (b)
0  0  0  1             1  1  1  1

0  0  0  1             0  0  0  1

0  0  0  1             0  0  0  1

0  0  0  1             0  0  0  1

(c)                    (d)
0  0  0  1             1  1  1  1

0  0  0  1             0  0  0  1

0  0  0  1             0  0  0  1

1  0  0  1             0  0  1  1
```

**Fig. 1.1**  Characters 1 and 7

**Table 1.1**  Row-major representation of patterns

| Figure label | 16-bit representation |
| --- | --- |
| (a) | 0 0 0 1 0 0 0 1 0 0 0 1 0 0 0 1 |
| (b) | 1 1 1 1 0 0 0 1 0 0 0 1 0 0 0 1 |
| (c) | 0 0 0 1 0 0 0 1 0 0 0 1 1 0 0 1 |
| (d) | 1 1 1 1 0 0 0 1 0 0 0 1 0 0 1 1 |

**Table 1.2** Hamming distance, the number of bits mismatched, between string pairs

| Pattern | (a) | (b) | (c) | (d) |
|---|---|---|---|---|
| (a) | 0 | 3 | 1 | 4 |
| (b) | 3 | 0 | 4 | 1 |
| (c) | 1 | 4 | 0 | 5 |
| (d) | 4 | 1 | 5 | 0 |

is the column label. The diagonal entries $(i = j)$ are all 0 as the Hamming distance between a pattern and itself is 0.

Note from Table 1.2 that patterns in Fig. 1.1a and c are similar with a minimum Hamming distance of 1 and similarly patterns in (b) and (d) are equally similar. So, digit 1 in (a) is similar to a noisy version of digit 1 in (c). A similar argument applies to digit 7 in (b) and (d).

It is possible to select a subset of features that might be adequate. For example, in Fig. 1.1a, b, c, and d if we consider *only the four bits in the first row* of the four characters, then the four bits in (a) and (c) are "0 0 0 1" revealing their similarity and similarly figures (b) and (d) share the same bit string "1 1 1 1" again indicating that (b) and (d) are similar in the 4-dimensional space. So, in this simple example case, values of the bits in the first row of the $4 \times 4$ characters are adequate to *discriminate* between the classes 1 and 7 here. This type of selection of a subset of features (bits here) from the given set (16 bits here) is generically called *feature selection*. Note that in this example, the 4 bits, out of 16, in the first row are *important (central)* and are *discriminative (diverse)*. Here, *centrality is specified by importance* of features and the values they assume across the the two classes of digits are *diverse and discriminatory*. Even though we have selected the first row here by visual inspection, it is possible to mechanically obtain such discriminating features using *mutual information (MI)* that we examine in a later chapter. It helps in ranking features.

Another scheme that could be used to reduce the number of features is *feature extraction*, where a possibly new set of features is obtained (extracted) from the given set. For example, by considering the column sums in the characters in figures (a), (b), (c), and (d) we get the vectors of their respective column sums to be $(0, 0, 0, 4)$ for (a); $(1, 1, 1, 4)$ for (b); $(1, 0, 0, 4)$ for (c); and $(1, 1, 2, 4)$ for (d). Even in this 4-dimensional case similarity between the pairs (a) and (c); and (b) and (d) can be seen based on Euclidean distance. In this case, the new features (column sums) extracted may be viewed as *linear combinations of the original features*. For example, if we consider the row-major ordering of the bits in the $4 \times 4$ binary patterns in Fig. 1.1 and number them as $b_1, \ldots, b_{16}$ where $b_1, b_2, b_3, b_4$ correspond to the first row, and so on, then the column sums may be viewed as

Sum of column 1 $= b_1 + b_5 + b_9 + b_{13}$
Sum of column 2 $= b_2 + b_6 + b_{10} + b_{14}$
Sum of column 3 $= b_3 + b_7 + b_{11} + b_{15}$
Sum of column 4 $= b_4 + b_8 + b_{12} + b_{16}$

**Fig. 1.2** Centroids of 1
and 7

| (i) | | | | (ii) | | | |
|---|---|---|---|---|---|---|---|
| 0 | 0 | 0 | 1 | 1 | 1 | 1 | 1 |
| 0 | 0 | 0 | 1 | 0 | 0 | 0 | 1 |
| 0 | 0 | 0 | 1 | 0 | 0 | 0 | 1 |
| 0.5 | 0 | 0 | 1 | 0 | 0 | 0.5 | 1 |

**Fig. 1.3** Character 1 with
diversity (**a**)–(**d**) and their
centroid (**e**)

| (a) | | | | (b) | | | | (e) | | | |
|---|---|---|---|---|---|---|---|---|---|---|---|
| 0 | 0 | 0 | 1 | 0 | 0 | 1 | 0 | 0.25 | 0.25 | 0.25 | 0.25 |
| 0 | 0 | 0 | 1 | 0 | 0 | 1 | 0 | 0.25 | 0.25 | 0.25 | 0.25 |
| 0 | 0 | 0 | 1 | 0 | 0 | 1 | 0 | 0.25 | 0.25 | 0.25 | 0.25 |
| 0 | 0 | 0 | 1 | 0 | 0 | 1 | 0 | 0.25 | 0.25 | 0.25 | 0.25 |

| (c) | | | | (d) | | | |
|---|---|---|---|---|---|---|---|
| 0 | 1 | 0 | 0 | 1 | 0 | 0 | 0 |
| 0 | 1 | 0 | 0 | 1 | 0 | 0 | 0 |
| 0 | 1 | 0 | 0 | 1 | 0 | 0 | 0 |
| 0 | 1 | 0 | 0 | 1 | 0 | 0 | 0 |

In each of the above sums, the bits in the respective column are weighted (multiplied by 1) and the rest of the 12 bits may be considered to have a zero weight each. Such a linear combination may be generically represented using $\sum_{i=1}^{16} w_i b_i$ where $w_i$ is the weight (multiplier) of the $i$th bit ($b_i$). For the sum of column 1, $w_1 = w_5 = w_9 = w_{13} = 1$ and the other weights are 0. There are popular schemes based on *principal component analysis* and others for feature extraction which we consider in detail later.

Feature selection and feature extraction are important *dimensionality reduction* techniques. We have briefly examined *linear feature extraction* in this section. *Non-linear feature extraction* also is gaining importance because of several applications including *embeddings in natural language processing and social networks*. We will examine the role of centrality and dimensionality in dimensionality reduction in later chapters.

## 1.3.2  Clustering and Classification

Centroid or sample mean of a collection of patterns in a cluster is a *central or representative* pattern of the cluster. It could also be viewed as a *representative or a prototype of a class*. Figure 1.2 depicts the centroid of patterns of digit 1 (i) and digit 7 (ii) in Fig. 1.2. Note that the 0.5 stands for the average of 0 and 1. Even though centroid is a popular representative of a cluster (class), it may not always work well. For example, Consider Fig. 1.3.

There are 4 patterns corresponding to digit 1 in (a), (b), (c), and (d) and their centroid is shown in (e). Here the centroid in (e) fails to "represent" the digit 1. Note that there is a sufficient diversity among the patterns in (a)–(d) and each corresponds to a 1 with a shift in the column having binary bits with value 1. In fact, each of them may be viewed as a representative of a subclass of 1s. So, one has to respect the *diversity* present in the class of patterns before capturing the *centrality*. It makes sense to represent each subclass with its respective centroid. In this context, the notions of centrality and diversity are different from the ones in the context of feature selection exemplified in the previous subsection. Here, *centrality* of a class/cluster of vectors is captured by their centroid and *diversity* could be present in various subclasses of a class. We will see later the role of centroid in clustering and classification. A more detailed discussion on *centrality and diversity in clustering and classification* will be undertaken in a later chapter.

## 1.3.3 Ranking

It involves ordering a collection of entities based on some specification. For example, in *machine learning* we might order a collection of features by using some ranking function which captures the discrimination capabilities of the features. In *social networks* we may order the nodes in the network based on their centrality or importance; a simple notion of centrality is characterized by the degree of the node. Similarly, in *information retrieval*, we may order the terms based on their frequency of occurrence in a given collection of documents. In *recommender systems*, we would like to order or rank the recommendations based on some notion of *importance or centrality*. Ranking is important in several of these tasks and some notion of *centrality* drives the ranking operation.

Ranking is the most essential and routinely performed operation by search engines. Search engines provide a ranked list of documents against a specified query. In the early days, search engines used to rank order the output list of documents by matching the content present in the query with that in each of the documents. So, a document is more central if its content matches better with that of the query. In a practical sense, combining the link structure across various web pages/documents along with their content to improve the search results. In this setting, a web page/document is more central if it matches in terms of its content and if the web page is central in its link structure with other web pages. Here, the notion of centrality comes through *PageRank*, the rank index of a web page based on its link structure. A web page has larger PageRank if larger PageRank web pages refer to it. So, in this combined context, a web page is central if it not only matches in content with the query but also has a good PageRank. So, notion of centrality is changed.

It is easy to see that ranking based on centrality alone may lead to a repetition of similar/identical documents appearing close to each other in the ranked list. This may not be desirable. So, some notion of *diversity* has to exploited. Keeping in mind that a typical user may navigate only through the first page of the ranked results output by a search engine, sufficient diversity needs to be shown in the document snippets shown on the first page; the other pages may not be visited by the user. So, enough diversity must be present in the early results. So, while ranking we need to consider both centrality and diversity among the displayed results.

One way to achieve this is to cluster the results rank ordered based on central-ity into cohesive groups, where each group has similar documents and documents from different clusters are dissimilar. To maintain diversity, one can select central documents from different clusters and display them on the first page. This way one can display on the first page documents that are diverse from each other and each document is central/important. This could be a generic framework that is useful in preserving diversity and centrality in different machine learning tasks.

### 1.3.4   Regression

Regression or curve fitting may be viewed as a generalized version of classification. In classification, we typically have a finite number of class labels; in regression we may view the number of classes as infinite. In regression, we need to learn a function $f$ which maps vectors such that

$$f : \mathcal{X} \to \mathcal{Y}, \text{ where } \mathcal{X} \subset \mathfrak{R}^d \text{ and } \mathcal{Y} \subset \mathfrak{R}^{d'}$$

The training set will consist of $n$ pairs of vectors $\{(x_i, y_i), i = 1, \ldots, n\}$ and the dimensionality, $d$, of $x \in \mathcal{X}$ could be large but the dimensionality, $d'$, of $y \in \mathcal{Y}$ is small or even one. So, in regression each $x$ can be mapped to one of potentially infinite (cardinality of $\mathfrak{R}$ with $d' = 1$) values (labels).

A major problem in classification and regression is *overfitting*. For example con-sider the following regression/curve fitting problem. Consider the pair of $x$ and $y$ values shown in Table 1.3 which are generated by using $y = f(x) = 1 + x + x^2$.

Suppose we consider the first 3 points in the table. It is possible to fit a degree-2 polynomial uniquely. So, if we start with a generic polynomial form given by

$$g_2(x) = a_0 + a_1 x + a_2 x^2$$

where $g_i(x)$ is a degree-i polynomial in $x$ and use the three points to fit a least-square fit, then we obtain the values of the coefficients to be $a_0 = 1$, $a_1 = 1$, and $a_2 = 1$ giving us back the polynomial that was behind generating the three pairs of values. The least-square fit error is 0 and $g_2(x) = f(x)$. Note that using $g_2(x)$, the fourth pair

**Table 1.3** Pairs of x and y values

| Pattern no | Value of x | Value of y = f(x) |
| --- | --- | --- |
| 1 | 0 | 1 |
| 2 | 1 | 3 |
| 3 | 2 | 7 |
| 4 | $\frac{1}{2}$ | $\frac{7}{4}$ |

also is correctly predicted because $g_2(\frac{1}{2}) = \frac{7}{4}$. It is obvious because the fourth point also is generated using the same function form.

However, it will be interesting to examine what happens if we fit a degree-1 or degree-3 polynomial using the same 3 pairs of points. Let us examine the degree-1 polynomial fit by using the least-square method. Let the generic form

$$g_1(x) = a_0 + a_1 x$$

be used to represent a degree-1 polynomial. In this case, the least-square fit gives us $a_0 = \frac{2}{3}$, and $a_1 = 3$ with an overall error value of 1 unit on the three points. So, $g_1(x) = \frac{2}{3} + 3x$ is the degree-1 polynomial obtained using the least-square fit. The form of $g_1$ is *simpler* than that of $g_2$ and it is a unique degree-1 polynomial.

Now let us consider fitting a degree-3 polynomial using the same 3 pairs of points. Let the generic form be

$$g_3(x) = a_0 + a_1 x + a_2 x^2 + a_3 x^3.$$

By using the least-square fit we may get several options. Let us choose one option given by $a_0 = 1$, $a_1 = \frac{3}{2}$, $a_2 = \frac{1}{4}$, and $a_3 = \frac{1}{4}$. So,

$$g_3(x) = 1 + \frac{3}{2}x + \frac{1}{4}x^2 + \frac{1}{4}x^3$$

and the error is 0 for this degree-3 fit also on the three points. However, if we consider the fourth pair of $x$ and $y$ values in the table given by $x = \frac{1}{2}$ and $y = \frac{7}{4}$, then $g_3(x)$ gives a value of $\frac{59}{32}$, instead of $\frac{7}{4}$ which is erroneous.

There are other degree-3 options in addition to the degenerate solution with $a_3 = 0$. For example, one such degree-3 polynomial, $g'_3(x)$, is given by

$$g'_3(x) = 1 + \frac{5}{3}x + \frac{1}{3}x^3.$$

Note that $g'_3(x)$ also makes a 0 error on the first three points and on the fourth value of $x$, it gives $g'_3(\frac{1}{2}) = \frac{45}{24}$, instead of $\frac{7}{4}$ leading to an error that varies from the error given by $g_3(x)$ on the fourth value of $x$.

This simple example illustrates the following generic behavior, which is called the *bias–variance dilemma*.

- Using some $d + 1$ pairs of $x$ and $y$ values generated from a degree-d polynomial, we can predict uniquely a generic degree-d polynomial using the least-square fit with 0 error.
- If we use these $d + 1$ pairs to predict a polynomial of degree less than $d$, then we will get a least-square fit with nonzero error. Such simpler models are typically erroneous and exhibit a smaller variance in the error. In the above example, there is a unique $g_1$ that minimizes the least-square error and so the variance in the

error is 0. Even in the presence of noise these simpler models are erroneous and are low-variance (in the error) fits. So, polynomials of degree $d$ or less are more *central* and lesser degree polynomials are said to *underfit* the data.

- However, if we use the same $d + 1$ points to fit a degree-D polynomial where $D$ is larger than $d$, then we may have multiple options and each option may give rise to a different error on additional points leading to a larger variance in the error even though the error values could be smaller leading to a low bias on these $d + 1$ points. So, these models exhibit high *diversity* and are said to *overfit* the data. These may be viewed as better *bias* controllers.
- The overall error may be viewed as a combination of a *bias term* and a *variance term* and these work in opposite directions. Overfitting models fit the data (along with noise if any) better and are bias-controllers and underfitting models exhibit higher bias and are variance-controllers.
- For example, the nearest neighbor classifier (NNC) is a nonlinear classifier and gives 100% accuracy on the training data. However, it may do badly on the test data. This is an example of overfitting. However, even in very high-dimensional spaces people use a linear SVM, instead of a kernel SVM, because of the simplicity of the former. So such a simpler, underfitting linear SVM may not exhibit 100% training accuracy, but it works very well on the test data leading to a smaller generalization error.

Bias–variance dilemma is important in both regression and classification. We encounter centrality and diversity in a different manner in the learning of the associated models. Simple or low-variance models may be viewed as central models and complex or flexible models overfit the data and exhibit diversity. The best model is obtained by controlling centrality and diversity appropriately.

### 1.3.5  Social Networks and Recommendation Systems

A social network is typically represented as a graph $G = < V, E >$. Here each social entity is represented as a node $v \in V$ in the graph and an edge $e_{ij} \in E$ represents the relation between the entities represented by nodes $i$ and $j$. For example, a friendship network will be represented by an undirected graph where each member of the network is represented by a node of the graph and an edge is present between nodes $i$ and $j$ if the corresponding individuals are friends. The graph is undirected because friendship is symmetric. One popular data structure that can be used to represent a graph is its *Adjacency Matrix A*. If there are $n$ nodes in the graph, then $A$ is an $n \times n$ matrix where $i$th row represents node $i$ and its $j$th column represents the $j$th node where $i, j \in \{1, 2, \ldots, n\}$. In a simple scenario, $A$ is a binary matrix and its $ij$th entry $a_{i,j}$ is 1 if nodes $i$ and $j$ have an edge in the graph else $a_{i,j} = 0$. We say that $i$ is a neighbor $j$ and vice versa if $a_{i,j} = 1$. Let $a_i$ be the $i$th row of $A$ for $i = 1, \ldots, n$; it is a row vector with $n$ entries corresponding to $n$ columns of $A$. *Centrality* is a well-defined and important notion in social networks. A product company might identify influential or central individuals in different communities and promote their product through these identified central individuals. Some popularly used centrality notions are:

- **Degree Centrality**: A node is central if its degree is large. This is the simplest notion of centrality.
- **Closeness Centrality**: A node is central if it is closer to a large number of nodes around it. Closeness could be based on the length of the path between nodes. Shorter the length of the path between a pair of nodes, the closer the two nodes.
- **Betweenness Centrality**: An edge is more central if it connects two or more important communities. For example, a single bridge between two or more geographical regions is crucial in maintaining connectivity between the regions. This notion is important in forming communities.
- **Eigenvector Centrality**: This is a recursive notion of centrality. A node is central if it is connected to central nodes. A nonzero vector $x$ is an eigenvector of $A$ if $Ax = \lambda x$ with $\lambda \in \Re$ being the eigenvalue of $A$ corresponding to $x$. Note that $x$ is a column vector of size $n$ and $x_i$ is the $i$th component of $x$, for $i = 1, \ldots, n$. So, from $Ax = \lambda x$ we get

$$\lambda x_i = \sum_{j=1}^{n} a_{i,j} x_j \Rightarrow x_i = \frac{1}{\lambda} \sum_{j=1}^{n} a_{i,j} x_j.$$

Entries in eigenvector $x$ indicate the centralities of the respective nodes; the above equation explicitly indicates the eigenvector centrality of node $i$, $x_i$, in terms of the eigenvector centralities of all the $n$ nodes, $x_1, \ldots, x_n$.

Note that even in this context also, centrality alone is not adequate. For example, if there are two central nodes $i$ and $j$, such that they have a large number of common neighbors, then it may not be meaningful to consider both of them for the promotional purposes as both of them may be able to influence the same set of nodes or nodes in the same community. In a such a case, it is important to be concerned about *diversity* among such central nodes also; for example, how diverse various central nodes are in terms of their influence on the other individuals or how nonoverlapping are their sets of neighbors are.

### 1.3.5.1  Recommender Systems

In addition to the link structure present in social networks, each node may also have some content associated with it. Such networks are called *information networks* where each node carries both the link information and content. In such networks, a dichotomy exists.

1. **Homogeneous Information Networks**: Here all the nodes are of the same type and edges are also of the same type. For example, in a citation network each

publication is represented by a node in the graph along with its associated content. A directed edge exists from node $i$ to node $j$ if the the publication represented by $i$ *refers to* the publication corresponding to node $j$. The corresponding graph is a directed graph in this example.

2. **Heterogeneous Information Networks**: There can be applications where we may need multiple types of nodes and also multiple types of edges. Such networks are called heterogeneous networks where either the nodes or edges are of different types. For example, in a *restaurant network* where we want to represent the (i) users, (ii) restaurants, (iii) cuisine, and (iv) their location information, then we end up with multiple types of nodes; a node may represent any one of these four types. Further, the edges also can be of multiple types. A user *visits* a restaurant; is a *friend of* another user; *stays in* a location; and *likes* Italian dishes. We may have many more types of relations in this network among types nodes other than *user*.

   Such heterogeneous information networks are exploited by *recommendation systems*. A restaurant is recommended to a user based on the notion of its importance/centrality that is characterized based on the preferences of the user. Even here *diversity* is important. For example, if two important/central restaurants are located in the same region and offer the same type of cuisine, then it may not be good to recommend both of them just based on their centrality. It may be meaningful to recommend to the user *diverse and central* restaurants.

## 1.4   Summary

In this chapter, we have introduced the notions of *centrality* and *diversity*. Also their roles in *search* in general and more specifically in important *machine learning tasks* including *representation*, *clustering and classification*, *ranking*, *regression* and application areas like *social networks* and *recommendation systems* are introduced. We will examine in more detail their role in specific tasks in later chapters.

## Bibliography

1. Murty MN, Devi VS (2011) Pattern recognition-an algorithmic approach. Undergraduate topics in computer science. Springer, London
2. Murphy KP (2012) Machine learning-a probabilistic perspective. MIT Press
3. Murty MN, Raghava R (2016) Support vector machines and perceptrons, Springer briefs in computer science. Springer, Cham
4. Bishop CM (2005) Neural networks for pattern recognition. Oxford University Press
5. Russell SJ, Norvig P (2015) Artificial intelligence: a modern approach. Pearson

# Chapter 2
# Searching

**Abstract** Search is the most basic and fundamental operation of computers. It plays a vital role in areas like artificial intelligence, machine learning, deep learning, and a variety of applications. We examine the role of search in several related topics in this chapter.

**Keywords** Exact match · Inexact match · Classification · Representation

## 2.1 Introduction

Matching is a fundamental operation in areas including artificial intelligence, machine learning, data mining, and pattern recognition. Different types of matching are *exact match* and *inexact match*. In the exact match, we insist that the matching item and the matched item are identical in some sense. In the inexact match, we look for similar items or approximately matching items.

### 2.1.1 Exact Match

The algorithms community has excelled in dealing with exact matching and currently, there is a growing interest in approximation algorithms. The problem in exact matching is as follows.

Given a collection of items $C_I$ containing $n$ elements and an item $x$, to find out whether $x$ is a member of $C_I$ or not. This problem is classically called the *search problem* and various algorithms for search are

1. *Linear Search*: This is also called the *sequential search* as $x$ is compared with the members of $C_I$ sequentially starting from the first element. The search stops and reports *success* if $x$ is matched with an element of $C_I$, else it reports *failure*. It requires $O(n)$ time if there are $n$ elements in $C_I$ in the worst case. That is why it is called linear search.

2. *Binary Search*: Here elements in $C_I$ are assumed to be in a sorted order. Let them be in nondecreasing order. Let the middle element of $C_I$, $C_I(mid)$, be $y$. First, $x$ is compared with $y$ and the search progresses as follows:

   a. If $x = y$ then stop and report *success*.
   b. If $x < y$ then recursively search for $x$ in $C_I(1)$ to $C_I(mid - 1)$.
   c. If $x > y$ then recursively search for $x$ in $C_I(mid + 1)$ to $C_I(n)$.

   It reports *failure* if $x$ fails to match with any element of $C_I$. This algorithm employs divide-and-conquer strategy and it is possible to show that it requires $O(log n)$ in the worst case. That is why it is also called *logarithmic search*.
3. *Hashing*: Here a function $h$, called hash function, is used to map items in $C_I$ to locations in an array. It is possible that the size of the array is smaller than $n$. In such a case, $h$ may map two different elements of $C_I$ to the same location in the array. Such a collision is resolved by maintaining a bucket containing all the elements that are mapped to the same location in the array. Then search for $x$ is carried out by matching elements in the bucket associated with $h(x)$. It is claimed that the search can be carried out in $O(1)$ or constant time.

Exact search where the matching operation outputs a collection of records that match the query/requirement is popular in databases. For example, from a collection of employee records of an organization, against a query *get names of all the employees whose salary is less than an amount $x$* get a collection of one or more employee records satisfying the query.

Even though database researchers concentrated typically on such exact matching, they have played an important role in making research in machine learning, specifically making information retrieval and search work. For example, database research has contributed more to the success of search engines, founding the area of data mining and also in the analysis of social and information networks.

### 2.1.2   Inexact Match

Typically in artificial intelligence, machine learning, and pattern recognition, we depend more on approximate or inexact search. Specifically, we depend on inexact search in representation, clustering, classification, working of search engines, and regression. We discuss it in detail next.

### 2.1.3   Representation

The nearest neighbor classifier (NNC) is the simplest and easy to comprehend classifier. In NNC, a set of training patterns

$$\mathcal{X} = \{(x_1, c_1), (x_2, c_2), \ldots, (x_n, c_n)\}$$

is given in which the $i$th pattern $x_i$ is labeled with $c_i$ for $i = 1, \ldots, n$. Here $c_i \in \{1, 2, \ldots, C\}$ for $i = 1, \ldots, n$ where $C$ is the number of classes. Using the NNC a test pattern $x$ is classified to belong to the same class as its nearest neighbor. So, if $x_j, j \in \{1, \ldots, n\}$ is the nearest neighbor of $x$, then the test pattern $x$ is assigned the label $c_j$ by NNC. Here the problem is one of making an inexact search where the nearest neighbor of $x$ is $x_j$ if

$$(x_j, c_j) = \arg\min_{(x_i, c_i) \in \mathcal{X}} d(x, x_i).$$

Here $d(p, q)$ is some distance, usually Euclidean distance, between the vectors $p$ and $q$.

A similar inexact search is used by the $K$- nearest neighbor classifier (KNNC) also. Here for a test pattern $x$, we find $K$ of its nearest neighbors from $\mathcal{X}$. Let the $K$ neighbors be $x^1, \ldots, x^K$ with their respective labels $c^1, \ldots, c^K$. We assign to $x$ the label that occurs with a maximum frequency (majority class label among the $K$ neighbors) among $c^1, \ldots, c^K$. So in the case of KNNC also we use *inexact search* to find the $K$ neighbors.

A *major problem with KNNC is that it does not work well in high-dimensional spaces* as $d(x, NN(x))$ may converge to $d(x, FN(x))$ as the dimensionality goes on increasing where $d(p, q)$ is the distance between patterns $p$ and $q$, $NN(x)$ is the nearest neighbor of $x$ and $FN(x)$ is the farthest neighbor of $x$. So, KNNC is not recommended for classifying documents or other such high-dimensional patterns. Some solutions to the problem are: (1) Feature Selection and/or (2) Feature Extraction.

In feature selection, we are given a collection of $n$ training patterns each represented as a vector in a $l$- dimensional space where the set, $F$, of features is $F = \{f_1, f_2, \ldots, f_l\}$ and the set, $\mathcal{C}$, of classes is $\mathcal{C} = \{c_1, \ldots, c_C\}$ are given as inputs. The output is $F'$, where $F' \subset F$ and $|F'| = d(< l)$. In other words, we select $d$ out of the $l$ features. There are different schemes for for this selection. One of the most popular schemes is based on *mutual information (MI)*. We can rank the $l$ features based on MI and select the top $d$ features. Mutual information between feature $f_i$ and class $c_j$ is given by

$$MI(f_i, c_j) = \sum_{i,j} p(I_i, I_j) log \frac{p(I_i, I_j)}{p(I_i).p(I_j)}$$

where we are assuming, for the sake of simplicity, that $f_i$ either occurs (value of $I_i = 1$) or does not occur (value $I_i = 0$); similarly, $I_j = 1$ if the class is $c_j$ and $I_j = 0$ if the class is non-$c_j$. MI gives us some kind of discriminative index of feature $f_i$ with respect to class $c_j$.

These probabilities can be computed as follows. Let $n$ be the number of training patterns; $n_{1,1}$ be the number of patterns in $c_j$ in which $f_i$ occurs; $n_{1,0}$ be the number of patterns in class non-$c_j$ (other than $c_j$) in which $f_i$ occurs; $n_{0,1}$ be the number of patterns in $c_j$ in which $f_i$ does not occur; and $n_{0,0}$ be the number of patterns in non-$c_j$ in which $f_i$ does not occur. So, $(n_{1,1} + n_{1,0})$ is the number of patterns in which $f_i$ occurs and $(n_{0,1} + n_{0,0})$ is the number of patterns in which $f_i$ does not occur. Similarly, $(n_{1,1} + n_{0,1})$ is the number of patterns in $c_j$ and $(n_{1,0} + n_{0,0})$ is the number of patterns in non-$c_j$. Then, we can interpret both $i$ and $j$ as binary variables by examining the following possibilities:

1. Feature $f_i$ is present in class $c_j \Rightarrow I_i = I_j = 1$. Consequently $p(I_i, I_j) = \frac{n_{1,1}}{n}$; $p(I_i = 1) = \frac{n_{1,0}+n_{1,1}}{n}$; $p(I_j = 1) = \frac{n_{0,1}+n_{1,1}}{n}$.
2. Feature $f_i$ is present in class non-$c_j \Rightarrow I_i = 1; I_j = 0$. So, $p(I_i, I_j) = \frac{n_{1,0}}{n}$; $p(I_i = 1) = \frac{n_{1,0}+n_{1,1}}{n}$; $p(I_j = 0) = \frac{n_{1,0}+n_{0,0}}{n}$.
3. Feature $f_i$ is absent in class $c_j \Rightarrow I_i = 0; I_j = 1$. So, $p(I_i, I_j) = \frac{n_{0,1}}{n}$; $p(I_i = 0) = \frac{n_{0,0}+n_{0,1}}{n}$; $p(I_j = 1) = \frac{n_{1,1}+n_{0,1}}{n}$.
4. Feature $f_i$ is absent in class non-$c_j \Rightarrow I_i = I_j = 0$. So, $p(I_i, I_j) = \frac{n_{0,0}}{n}$; $p(I_i = 0) = \frac{n_{0,0}+n_{0,1}}{n}$; $p(I_j = 0) = \frac{n_{1,0}+n_{0,0}}{n}$.

Once we have each of the $l$ features with the corresponding $MI$ value, we can rank them based on the $MI$ value. This ranking is based on some kind of *discriminative evidence* in the form of $MI$ values of the features. Thus, $MI$ plays the role importance/centrality of the feature. It is adequate to select the top $d$ features based on the $MI$ values to represent data items in a *binary classification* problem, where there are only two classes. However, if there are more than two classes, that is $C > 2$, then top $d$ features may not be adequate; among the central/important features we need to select $d$ *discriminative/diverse* features. We need to select $d$ features such that collectively they can discriminate between all the $C$ classes; it is not good to select a larger number of features that can discriminate a class $c_j$ from the rest at the cost of some of the other (non-$c_j$) classes. So, such a selection involves both centrality and diversity.

In feature extraction, we select $d$ new features $f_1', \ldots, f_d'$ that are linear or nonlinear combinations of the given $l$ features; For example, if the extracted features are linear combinations, then

$$f_j' = \sum_{i=1}^{l} \alpha_{i,j} f_i,$$

where $\alpha_{i,j}$ is the importance/weight of $f_i$ for $f_j'$. It is important that $f_j'$ is as diverse as possible from $f_k'$, $k \neq j$. Principal Components (PCs) are the eigenvectors of the sample covariance matrix of the data set; they are ranked based on the ordering of the eigenvalues of the matrix in a nonincreasing order. The first $PC$ is the eigenvector corresponding to the largest eigenvalue while the last $PC$ is the eigenvector pairing up with the smallest eigenvalue. We need to select the best $d$ principal components that can discriminate patterns from different classes. It is well-known that the $PCs$ are orthogonal to each other as the covariance matrix is symmetric. It is possible that the first $d$ *PCs* are not the best features for discriminating between different classes. However, the best $d$ *PCs* show diversity by being orthogonal to each other.

There are other applications of search in classification. For example, searching for a classification model based on some optimality criterion.

## 2.2 Proximity

A cluster or a class of points satisfies some kind of similarity or proximity. This proximity could be captured based on some common properties of the data points in the cluster/class. This *gestalt* property is captured by

- *Probability distribution*: The points in a class/cluster are drawn from an unknown underlying probability distribution. This view permits us to characterize proximity between a pair of classes/clusters. For example, *Kullback−Leibler* divergence (*KLD*) between a pair of classes/clusters as

$$KLD(p_i, p_j) = \sum_{k=1}^{l} p_{i,k} log\left(\frac{p_{i,k}}{p_{j,k}}\right)$$

  where $p_i$ is the probability mass function underlying class $c_i$ in a $l$-dimensional space.
- *Vector space*: The vectors in a class/cluster are elements of a $l$-dimensional vector space. So, a collection of patterns can be viewed as a matrix $A_{n \times l}$. Typically, the vector space is $\Re^l$. In such a case, matrix $A$ could be factorized into $B_{n \times K}$ and $C_{K \times l}$ where $K$ is the number of soft clusters. Here, proximity is implicit and is characterized by the semantics underlying the matrix factorization, $A = BC$.
- *Logical Description*: A cluster/class is described by a *concept* that is a formula in a formal language like mathematical logic. A collection of objects is described by the concept based on the properties shared by the objects in it. Here, inference based on logic could be used to check whether a pattern is from a class/cluster or not.

In statistical pattern recognition, machine learning, and deep learning, it is assumed that patterns are vectors and proximity between pairs of patterns is used to extract the clusters. The notion of proximity is used in both clustering and classification.

### 2.2.1 Distance Function

Distance between patterns $X_i$ and $X_j$ is denoted by $d(X_i, X_j)$ and the most generic distance measure is the *Minkowski Distance Metric* and it is given by

$$d(X_i, X_j) = \left( \sum_{k=1}^{l} |x_{ik} - x_{jk}|^p \right)^{\frac{1}{p}}$$

A pair of patterns are closer or similar if the distance between them is smaller. Different positive integer values of $p$ lead to different metrics. The most popular among them is the Euclidean distance which corresponds to $p = 2$. Even though metrics are useful in terms of computational requirements, they are not essential in *machine learning*.

For example, *KLDivergence* is not a metric; still, it is popularly used in several important applications. Proximity is characterized by either distance or similarity.

### 2.2.1.1  Cosine Similarity

Is popularly used in information retrieval and is defined as

$$cos(X_i, X_j) = \frac{\sum_{k=1}^{l} x_{ik} x_{jk}}{|| X_i || \ || X_j ||}$$

## 2.2.2  Clustering

In partitional clustering, we search for an optimal partition based on some criterion function. For example, the popular $K$-means algorithm (KMA) is given in Algorithm 1. In clustering applications, *KMA* is popularly used as it is a linear time algorithm requiring $O(Kn)$ time. The $K$ centroids are important to represent the underlying clustering structure or the $K$-partition; *KMA* generates a hard partition of the data set based on minimizing the *squared error criterion (SEC)* which is given by

$$SEC(\{c_1, \ldots, c_K\}) = \sum_{i=1}^{K} \sum_{x \in c_i} d^2(x, \mu_i),$$

where $d^2(x, \mu_i)$ is the squared Euclidean distance or deviation of $x \in c_i$ from the corresponding centroid $\mu_i$. The algorithm may end up in a local optimum of *SEC* if the initial centroids/representatives are not selected properly. We illustrate it using a 2-d example.

*Example 2.1* Consider a collection of 2-dimensional patterns given by

$$\mathcal{X} = \{(1, 1)^t, (1, 2)^t, (2, 1)^t, (2, 2)^t, (7, 2)^t, (7, 7)^t, (8, 2)^t, (8, 7)^t, (9, 7)^t, (8, 8)^t\}$$

---

**Algorithm 1** $K$-Means Algorithm

---

**Input:** A $d$-dimensional data set $\mathcal{X}$ with $n$ data objects and values for parameters $K$.
**Output:** A hard partition of $\mathcal{X}$.
1: Select $K$ $(< n)$ data points as initial representatives/centers of $K$ clusters, each representing a cluster. Let them be $x^1, \ldots, x^K$.
2: Let $x$ be a point of the set $X_R = \mathcal{X} - \{x^1, \ldots, x^K\}$. Assign $x$ to cluster $c_i$ if

$$d(x, x^i) < d(x, x^j) \text{ for } j \neq i,$$

where $d(x, y)$ is the Euclidean distance between $x$ and $y$. Assign each of the points in $X_R$ to its nearest cluster based on minimum Euclidean between the point and the cluster representative.
3: Compute the sample mean, $\mu_i$, of cluster $c_i$, where $\mu_i = \frac{1}{|c_i|} \sum_{x \in c_i} x$.
4: Assign each element of $\mathcal{X}$ to its nearest cluster, based on the Euclidean distance between the point and $\mu_i$s. Stop if there is no change in the assignment of points to clusters over two successive iterations. Else go to step (3).

---

Let us consider application of *KMA* on this data set with $K = 3$ based on two different selection of the initial centroids as given below;

1. Let the three initial cluster representatives for $c_1, c_2$, *and* $c_3$ be $(1, 1)^t$, $(1, 2)^t$, *and* $(2, 2)^t$, respectively. So, the assignment of the points to their respective clusters during iterations of KMA (steps 3 and 4 of *KMA*) is as shown in Table 2.1.
2. Instead, if we take the three initial representatives to be $(2, 2)^t$, $(7, 2)^t$, *and* $(7, 7)^t$ for $c_1, c_2, c_3$ respectively, then the assignment is going to be different and is given in Table 2.2.

It is interesting to analyze the partitions shown in Tables 2.1 and 2.2. Some important observations are the following.

**Table 2.1** Iterations of *KMA*: The assignment is same across iterations 2 and 3. So, *KMA* stops at the end of iteration 3 (step 4 of *KMA*)

| Iteration | Assignment | $c_1$ | $c_2$ | $c_3$ |
|---|---|---|---|---|
| | Representative | $(1, 1)^t$ | $(1, 2)^t$ | $(2, 2)^t$ |
| 1 | Other points | $(2, 1)^t$ | $\ldots$ | $(7, 2)^t, (8, 2)^t, (7, 7)^t$ $(8, 7)^t, (9, 7)^t, (8, 8)^t$ |
| | Centroid | $(1.5, 1)^t$ | $(1, 2)^t$ | $(7, 5)^t$ |
| 2 | All 9 Points | $(1, 1)^t$ | $(1, 2)^t$ | $(7, 2)^t, (8, 2)^t, (7, 7)^t$ |
| | | $(2, 1)^t$ | $(2, 2)^t$ | $(8, 7)^t, (9, 7)^t, (8, 8)^t$ |
| | Centroid | $(1.5, 1)^t$ | $(1.5, 2)^t$ | $(7.9, 5.5)^t$ |
| 3 | All 9 Points | $(1, 1)^t$ | $(1, 2)^t$ | $(7, 2)^t, (8, 2)^t, (7, 7)^t$ |
| | | $(2, 1)^t$ | $(2, 2)^t$ | $(8, 7)^t, (9, 7)^t, (8, 8)^t$ |
| | Centroid | $(1.5, 1)^t$ | $(1.5, 2)^t$ | $(7.9, 5.5)^t$ |
| | Contribution to SEC | 0.5 | 0.5 | 46.36 |
| | SEC of the partition | **47.36** | | |

**Table 2.2** Iterations of *KMA*: The assignment is same across iterations 1 and 2. So, *KMA* stops at the end of iteration 2 (step 4 of *KMA*)

| Iteration | Assignment | $c_1$ | $c_2$ | $c_3$ |
|---|---|---|---|---|
|  | Representative | $(2, 2)^t$ | $(7, 2)^t$ | $(7, 7)^t$ |
| 1 | Other points | $(1, 1)^t, (1, 2)^t$ $(2, 1)^t$ | $(8, 2)^t$ | $(8, 8)^t, (8, 7)^t$ $(9, 7)^t$ |
|  | Centroid | $(1.5, 1.5)^t$ | $(7.5, 2)^t$ | $(8, 7.25)^t$ |
| 2 | All 9 Points | $(1, 1)^t, (1, 2)^t$ $(2, 1)^t, (2, 2)^t$ | $(7, 2)^t$ $(8, 2)^t$ | $(7, 7)^t, (8, 7)^t$ $(9, 7)^t, (8, 8)^t$ |
|  | Centroid | $(1.5, 1.5)^t$ | $(7.5, 2)^t$ | $(8, 7.25)^t$ |
|  | Contribution to SEC | 2.0 | 0.5 | 2.75 |
|  | SEC of the partition | **5.25** | | |

1. In this simple example, the 3-partition characterized by Table 2.2 is the best in terms of *SEC* value of 5.25. It is better than the 3-partition shown in Table 2.1 as the corresponding *SEC* value is 47.36.
2. The partition, shown in Table 2.1, based on selecting the initial cluster representatives that are very close to each other resulted in a local optimum of *SEC*.
3. On the other hand, the partition obtained using initial representatives that are far apart from each other (*diverse*) gave the best *SEC* value.

So, characterizing clusters by their centroids/central representatives is *not adequate*. We need to capture the diversity present in the cluster structure by *selecting initial representatives that exhibit diversity*. *K*-Means++ is a randomized version of *KMA* that addresses this problem by exploiting diversity in the selection. However, if we give too much importance to diversity, then outlier patterns might be selected as initial cluster representatives. So, one has to ensure that each such selected representative is central and comes from a dense region.

### 2.2.3  Classification

In classification, we learn a classification model using a set of training patterns. Test patterns or patterns that are not yet labeled are labeled or classified using the classification model. Some of the important issues in building classifiers include the following.

#### 2.2.3.1 Representation

For the machine to learn the model, typically we need to represent the patterns/data points as vectors. We use discriminative or central features in vector representation; further one needs to select these features respecting the diversity as mentioned earlier.

Not only representing each pattern as a vector, in some cases we need to search for a compact representation of the training dataset. For example, if there are $n$ training patterns, then *exemplar-based* classifiers like *NNC* will require $\mathcal{O}(n)$ time to compute the nearest neighbor of each test pattern. One way to reduce the effort is by condensing or reducing the dataset. The condensed nearest neighbor classifier (*CNNC*) condenses the dataset size from $n$ to $m$. CNNC works by first obtaining the condensed dataset and then use it for classifying a test pattern by using *NNC* and the condensed set. We explain it using Algorithm 2.

---

**Algorithm 2** Condensation

---

**Input:** A labeled training data set $\mathcal{X} = \{(x_1, lab_1), (x_2, lab_2), \ldots, (x_n, lab_n)\}$ with $n$ data objects; $lab_i$ is the *class label* of $x_i$.
**Output:** A condensed dataset, *CONDENSED* that is a subset of $\mathcal{X}$.
1: *CONDENSED* = $\{(x_1, lab_1)\}$. *REMAIN* = $\mathcal{X}$ − *CONDENSED*.
2: If *REMAIN* = $\phi$ then goto step 5.
3: Let $(x, lab)$ be the first pattern in *REMAIN*. Let $(x', lab') \in$ *CONDENSED* be such that $x'$ is the nearest neighbor (NN) of $x$.
4: If $lab \neq lab'$ then *CONDENSED* = *CONDENSED* $\cup \{(x, lab)\}$. Let *REMAIN* = *REMAIN* − $\{(x, lab)\}$. Goto step 2.
5: Stop if *CONDENSED* is not updated during the entire iteration. Otherwise,

$$REMAIN = \mathcal{X} - CONDENSED$$

. Goto step 2.

---

Once we get the set *CONDENSED*, a condensed set of training patterns, we use it along with *NNC* to classify the test patterns.

*Example 2.2* Consider a collection of training patterns. These are 2-dimensional patterns belonging to three classes given by

$$\mathcal{X} = \{((1, 1)^t, 1), ((2, 2)^t, 1), ((3, 2)^t, 1), ((5, 2)^t, 2), ((4, 2)^t, 2), ((7, 7)^t, 3), ((8, 7)^t, 3)\}$$

The generation of *CONDENSED* set using Algorithm 2 is illustrated using Table 2.3.

If we consider the same dataset in the order $\mathcal{X} = \{((1, 1)^t, 1), ((2, 2)^t, 1), ((3, 2)^t, 1), ((4, 2)^t, 2), ((5, 2)^t, 2), ((7, 7)^t, 3), ((8, 7)^t, 3)\}$ then we have the *CONDENSED* updated as shown in Table 2.4.

So, even though it condenses the dataset to retain some *essential/central data points*, it is *order dependent* as it processes the data points incrementally. Each point is processed in isolation based on *NNC* and *CONDENSED*. Different orders of the

**Table 2.3** Stepwise processing of condensation

| Step number | CONDENSED | REMAIN |
|---|---|---|
| 1 | $\{((1, 1)^t, 1)\}$ | Set of last 6 patterns in $\mathcal{X}$ |
| Next 2 patterns | No change | Set of last 4 patterns in $\mathcal{X}$ |
| 4 | $\{((1, 1)^t, 1), ((5, 2)^t, 2)\}$ | Set of last 3 patterns in $\mathcal{X}$ |
| 4 | $\{((1, 1)^t, 1), ((5, 2)^t, 2), ((7, 7)^t, 3)\}$ | $\{((8, 7)^t, 3)\}$ |
| 5 | $\{((1, 1)^t, 1), ((5, 2)^t, 2), ((7, 7)^t, 3)\}$ | $\{((2, 2)^t, 1), ((3, 2)^t, 1), ((4, 2)^t, 2), ((8, 7)^t, 3)\}$ |
| 5 | $\{((1, 1)^t, 1), ((5, 2)^t, 2), ((7, 7)^t, 3)\}$ | $\phi$ |

**Table 2.4** Stepwise processing of condensation with a different ordering

| Step number | CONDENSED | REMAIN |
|---|---|---|
| 1 | $\{((1, 1)^t, 1)\}$ | Set of last 6 patterns in $\mathcal{X}$ |
| Next 2 patterns | No change | Set of last 4 patterns in $\mathcal{X}$ |
| 4 | $\{((1, 1)^t, 1), ((4, 2)^t, 2)\}$ | Set of last 3 patterns in $\mathcal{X}$ |
| 4 | $\{((1, 1)^t, 1), ((4, 2)^t, 2), ((7, 7)^t, 3)\}$ | $\{((8, 7)^t, 3)\}$ |
| 5 | $\{((1, 1)^t, 1), ((4, 2)^t, 2), ((7, 7)^t, 3)\}$ | $\{((2, 2)^t, 1), ((3, 2)^t, 1), ((5, 2)^t, 2), ((8, 7)^t, 3)\}$ |
| 4 | $\{((1, 1)^t, 1), ((4, 2)^t, 2), ((7, 7)^t, 3), ((3, 2)^t, 1)\}$ | $\{((5, 2)^t, 2), ((8, 7)^t, 3)\}$ |
| 5 | $\{((1, 1)^t, 1), ((4, 2)^t, 2), ((7, 7)^t, 3), ((3, 2)^t, 1)\}$ | $\phi$ |

input data might give rise to different outputs as shown in Table 2.3 and 2.4. One solution to this problem of order dependence is to modify the CNNC by introducing *diversity* as shown in Algorithm 3

---

**Algorithm 3** Modified Condensed Set

---

**Input:** A labeled training data set $\mathcal{X} = \{(x_1, lab_1), (x_2, lab_2), \dots, (x_n, lab_n)\}$ with $n$ data objects; $lab_i$ is the *class label* of $x_i$.

**Output:** A condensed dataset, *CONDENSED* that is a subset of $\mathcal{X}$.

1: $MCONDENSED = \{(x^1, 1), (x^2, 2), \dots, (x^C, C)\}$, where $(x^i, i)$ is a typical/representative pattern from class $i$, $i = 1, \dots, C$. $REMAIN = \mathcal{X} - MCONDENSED$.

2: If $REMAIN = \phi$ then goto step 5.

3: Let $(x, lab)$ be the first pattern in $REMAIN$. Let $(x', lab') \in MCONDENSED$ be such that $x'$ is the nearest neighbor (NN) of $x$.

4: If $lab \neq lab'$ then $MCONDENSED = MCONDENSED \cup \{(x, lab)\}$. Let $REMAIN = REMAIN - \{(x, lab)\}$. Goto step 2.

5: Stop if $MCONDENSED$ is not updated during the entire iteration. Otherwise,

$$REMAIN = \mathcal{X} - MCONDENSED$$

. Goto step 2.

---

**Table 2.5** Typical patterns

| Class | Patterns | Centroid | Typical (nearest to centroid) |
|-------|----------|----------|-------------------------------|
| 1 | $(1, 1)^t, (2, 2)^t, (3, 2)^t$ | $(2, 1.66)^t$ | $(2, 2)^t$ |
| 2 | $(5, 2)^t, (4, 2)^t$ | $(4.5, 2)^t$ | $(5, 2)^t$ |
| 3 | $(7, 7)^t, (8, 7)^t$ | $(7, 5, 7)^t$ | $(7, 7)^t$ |

In step (1), we add to *MCONDENSED* one typical pattern from each class. Such a typical pattern could be obtained by considering a pattern nearest to the centroid of the class. For example, if we consider the data set $\mathcal{X} = \{((1, 1)^t, 1), ((2, 2)^t, 1), ((3, 2)^t, 1), ((5, 2)^t, 2), ((4, 2)^t, 2), ((7, 7)^t, 3), ((8, 7)^t, 3)\}$, then typical patterns in the three classes are indicated in Table 2.5.

Note that $MCONDENSED = \{((2, 2)^t, 1), ((5, 2)^t, 2), ((7, 7)^t, 3)\}$ by including all the three typical patterns and there are no updates as all the remaining patterns are correctly classified. Further, these typical patterns are both central and also account for diversity by selecting core/close to centroid patterns from each of the classes. This modification can help in improving *order independence*.

Centrality and diversity are important, perhaps in a different form, in some of the other classifier learning tasks including *model selection*. If we select a *simpler model*, then it can have less variance and more bias as a single simple model may fail to capture the diversity or details in the data; such a simple model is more central because of its simplicity. Different such simple models may emerge because each of them may see the diversity in the data or details in the data differently; so, diversity among the simple models is because of the inherent bias in the data. Naturally, we need to control both centrality and diversity. We will examine in a more detailed manner in a later chapter.

### 2.2.4 Information Retrieval

Searching is the most essential and routinely required operation in *information retrieval*. Search engines or retrieval systems provide a ranked list of documents against a user-specified query. In the early days, search engines used to rank-order the output list of documents by matching the content present in the query with that in each of the documents. So, a document is more central if its content matches better with that of the query.

It is typically assumed that underlying a collection of documents, $\mathcal{D}$ there is a vocabulary, $V$, of words/phrases that occur in the documents. So, in a simple binary representation of documents it is possible to view each document as a binary vector of size $|V| = l$ where the th bit is 1 if the th term in the vocabulary has occurred in the document, else the th bit is 0, where $i = 1, \ldots, l$. Here, a user query also is represented as a $l$-bit binary vector like the documents. In this case, the matching between a query $Q$ and a document $d \in \mathcal{D}$ is based on *Boolean search*. We say that $d$

satisfies or is *relevant* to $Q$ if all the terms in $Q$ are present in $d$ also when the query is viewed as an *and query* (a conjunction of terms in it).

A more popular representation of a document or a query is based on $TF - IDF$, a product of the *term frequency* ($TF$) and the *inverse document frequency* ($IDF$). Let

$$\mathcal{D} = \{d_1, d_2, \ldots, d_n\}$$

be a collection of $n$ documents and let

$$V = \{t_1, t_2, \ldots, t_l\}$$

be the vocabulary of $l$ terms (words/phrases). Let $TF_{ij}$ be the frequency of term $t_j$ in document $d_i$. Let $IDF_j$ be

$$IDF_j = log(\frac{n}{n_j})$$

be a simple characterization of the inverse document frequency, where $n$ is the number of documents and $n_j$ is the number of documents in which term $t_j$ occurred. A frequent term like "is" or "the" occurs in all the $n$ documents ($n_j = n$ for such terms) and so $IDF_j = log(1) = 0$, even though $TF_{ij}$ can be large. On the other hand rare terms like "abracadabra" may occur in a small number of documents. If $n_j = 1$, then $IDF_j = log(n)$ which could be large; however value of $TF_{ij}$ will be very small (0 or close to being 0). The term $TF - IDF$ is defined as

$$TF - IDF(i, j) = TF_{ij} \times IDF_j.$$

Note that $TF - IDF(i, j)$ will be very small whether $t_j$ is frequent or rare. Terms with medium frequency will have larger $TF - IDF$ values.

Each document is represented as an $l$-dimensional numerical vector, one number per term and this number, for term $t_j$ in document $d_i$ is $TF - IDF(i, j)$ for $j = 1, \ldots, l$ and $i = 1, \ldots, n$. In a similar way, the query $Q$ also is represented as a $l$-dimensional vector based on the respective $TF - IDF$ values. Similarity between $Q$ and a document $d$ is obtained by typically using the *cosine of the angle* between $d$ and $Q$ given by

$$sim(Q, d) = \frac{Q^t d}{||Q|| ||d||}.$$

A document $d_i$ is ranked better than another document $d_j$ if $sim(Q, d_i) > sim(Q, d_j)$.

However, in applications where the link structure is also available, it is advantageous to combine information from both the content and structure. For example, in web data mining the link structure across various web pages/documents along with their content can improve the search results. In this setting a web page/document is more central if it matches the query in terms of its content and if different web pages are equally important/central to the query, then diversity among them could be established through their importance through the link structure. Here, a web page/document matching the query, in terms of content, is given a better rank if it is important through its link structure also. A web page has a larger PageRank if larger PageRank web pages refer to it. So, in this combined context, a web page is treated as more relevant to the query if it not only matches in content with the query but also has a good PageRank. So, notion of centrality is changed.

### 2.2.5  Problem Solving in Artificial Intelligence (AI)

Search plays an important role in *problem solving*. Typically, a problem is solved by representing different solution paths by using transitions across states describing different configurations. Even though this is a generic setting that can deal with varieties of problems including playing games, solving puzzles, optimization, we illustrate the underlying ideas using a simple example puzzle.

Let us consider a *3-puzzle* problem shown in Fig. 2.1 where we are given some *initial configuration/state* of the puzzle and would like to reach some *specified/goal state*. The top left configuration in the figure is the initial state and the top right one is the goal state. Each configuration/state has 3 tiles that are numbered 1, 2, *or* 3 and one of the locations is free so that the tiles can move appropriately. It is convenient to abstract each possible move by viewing as to how the free space is moving. In the case of the Initial State in the figure, free space can be occupied by either the tile numbered 3 or tile numbered 2. So, equivalently the free space can either move right (**R**) or down (**D**). The resulting states are shown as left and right children nodes of the root node (initial state). From the right child by moving the free space right (**R**) we get the goal state as shown in the figure.

Here, we need to reach the goal state by using a sequence of *legal operations*. These are

1. The free space can move either horizontally or vertically; it can not move diagonally.

**Fig. 2.1**  State transitions in 3-puzzle problem

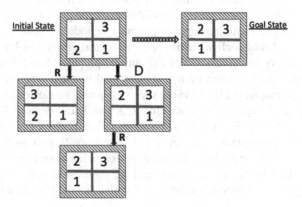

2. One can move the free space only by one position at a time. For example, from the initial state we cannot move the free space to reach the goal state by moving it by a sequence of two or more operations at a time.

In addition, we can have other problem-specific constraints like from the right child of the root node we can not move the free space up to go back to its parent node. We can record all the generated states so far to deal with this need by avoiding generation of the same state multiple times. So, there will not be any cycles in the resulting structure and also each state will have a single parent. This will ensure that the resulting structure is a tree.

This activity may be viewed as searching the set of states till the goal state is reached. Such a search is conducted by considering all possible legal operations on each state starting from the initial state. Such a transition from a state to another is captured by parent and child nodes, respectively. The entire search process can be abstracted by a tree under the specified constraints. Two popular schemes for search are *breadth-first search* (*BFS*) and *depth-first search* (*DFS*).

In *BFS*, we generate and examine the nodes in a level-wise manner; we consider all the children nodes of the root, one after the other, before proceeding further. For example, in Fig. 2.1, the left child of the root node has to be expanded before the right child. On the contrary, in *DFS*, we generate nodes along a path from the root and examine whether the goal state is reached in which case the process stops. If the goal node is not reached after a prespecified depth, then it will explore a different path from the root.

A more efficient and focused search is based on employing a *heuristic*. Such a search is called *heuristic search* or *informed search* as the heuristic employs some problem-specific knowledge. For example, we may consider a node that is more *similar* to the goal node before considering nodes that are less similar. Let the similarity between the goal and the current configurations be the number of rows matching between the two. So, in the example shown in Fig. 2.1, the root node is considered first. There are two children that are possible as shown in the figure. Among these two, the right child is more similar to the goal node as the first row matches between the two whereas there is no row matching between the goal and the left child. So, the right child is considered before the left child of the root as shown in Fig. 2.1. The path from the root (initial state) to the goal state is of length 2. The goal state is reached with the minimum number of node expansions that is 2 here.

Using such heuristics helps in several problem-solving situations. However, not every heuristic satisfies the optimal property. Some heuristics may even miss the optimal solution even if it exists and in some other cases they may mislead. For example, in Fig. 2.1 every state shown may be viewed as having the sequence 1, 2, 3 or its cyclic permutations like 2, 3, 1 or 3, 1, 2 ignoring the free space. Note that it is not possible to reach a sequence 2, 1, 3 or its cyclic permutations like 1, 3, 2 or 3, 2, 1 from a state with sequence 1, 2, 3 or its cyclic permutations as shown in Fig. 2.2.

Note that a heuristic that looks for match among the columns might assign a score of 1 to the similarity between the initial state and the goal state. However, using the legal operations considered earlier, it is not possible to reach the goal state from

**Fig. 2.2** Impossible
3-Puzzle Problem

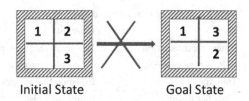

Initial State  Goal State

the initial state shown in Fig. 2.2. So, some important properties of these search algorithms are

1. An uninformed search like *BFS* guarantee reaching the goal if it is possible to reach it from the initial state. This is the *completeness property* of a search algorithm. However, it may not provide an optimal solution, for example, in terms of node evaluations. In the problem specified in Fig. 2.1 *BFS* needs to expand 3 nodes before encountering the goal state.
2. So, an uninformed search *may not be optimal*. Note that in Fig. 2.1 it is possible to reach the goal state by expanding just 2 nodes as discussed earlier.
3. On the other hand, heuristic search can speed up the problem-solving process by combining a complete search like BFS with the knowledge of the problem in the form of an appropriate heuristic.
4. However, it is possible that heuristic search may fail to reach the goal state even when a legal path from the initial state to the goal state exists. This is called the *admissibility property* of a search algorithm.
5. Heuristic search also might fail to detect that it is not possible to reach the goal state from some initial state as in the case discussed in Fig. 2.2.

So, complete search based algorithms may not be optimal; they capture the *diversity* present in the space of solutions or sequences of paths. Heuristic search, on the other hand, provides optimal search perhaps *at the cost of admissibility*. So, heuristic search is focused/centrality preserving search. The well-known algorithm $A*$, which we do not discuss here, judiciously combines the heuristic power with some uninformed search to provide an optimal/admissible algorithm.

## 2.3 Summary

In this chapter, we have considered the role of search in a variety of important tasks in AI, machine learning, representation, and information retrieval. The specific roles of search are that we search for:

- A *set of features* to represent patterns.
- A *set of representative/centroid patterns* in partitional clustering.
- An *appropriate subset of training patterns* for efficient classification.
- A *ranked collection of documents/snippets* against a user's query in information retrieval.

**Table 2.6**  Roles of centrality and diversity

| Task | Centrality | Diversity |
|---|---|---|
| Pattern representation | Discriminating features | Discriminate all classes |
| $K$-Prototype clustering | Centroids/Prototypes | Far apart representatives |
| Training | Essential training points | Represent intra-class diversity |
| Classifier learning | Simple models | Data variability |
| Search engine | Pagerank | Content match |
| Search in AI | Heuristic search | Complete search |
| Regression | Low variance | Bias preserving |
| Recommendations | Based on user | Diverse recommendations |
| Social networks | Centrality of entities | Multiple types of nodes |

- An optimal/admissible scheme in *problem solving in AI.*

In the previous chapter also, we discussed how centrality and diversity are useful in search in different tasks. A summary of different roles played by centrality and diversity is provided in Table 2.6.

# Bibliography

1. Nilsson NJ (1998) Artificial intelligence: a new synthesis. Morgan Kaufmann
2. Hastie T, Tibshirani R, Friedman J (2001) The elements of statistical learning. Springer, New York
3. Knuth DE (1998) The art of computer programming: sorting and searching, vol 3. Addison-Wesley
4. Manning CD, Schütze H, Raghavan P (2009) An introduction to information retrieval. Cambridge University Press
5. Nandanwar S, Moroney A, Murty MN (2018) Fusing diversity in recommendations in heterogeneous information networks. In: Proceedings of WSDM, 5–9 Feb 2018. Los Angeles, California, USA

# Chapter 3
# Representation

**Abstract** Representation is important in machine-based pattern recognition, AI, and machine learning. We need to represent states and state transitions appropriately in AI-based problem-solving. Similarly, in clustering and classification, we need to represent the data points, clusters, and classes.

**Keywords** Representation · Classification · Clustering · AI

## 3.1 Introduction

In dealing with several tasks in artificial intelligence and machine learning, we need to represent the problem/data appropriately so as to capture the knowledge about the problem/application. Each application might have a different requirement for the representation. We consider some of the relevant representations next.

## 3.2 Problem Solving in AI

Recall the description of the 3-puzzle problem discussed in the previous chapter. Each configuration/state is represented as a node in a tree/graph. This would help in implementing *BFS* and *DFS* directly on the nodes of the tree/graph. Note that searching on graphs can be complex compared to searching on trees. This helps in directly importing the well-established properties of these search schemes into the current problem-solving context.

Consider the puzzle problem shown in Fig. 3.1. The initial and goal state could be represented as a node in a graph as discussed in the previous chapter. Instead, it is also possible to represent each state as a string. For example, the initial state in the figure may be represented by 1, 2, #, 3 where the tile numbers are presented row-wise and the free space is represented by #. The given problem in the figure may be represented as

$$(\textit{Initial State}) \ 1, 2, \#, 3 \ \rightarrow^{to} \ 2, \#, 1, 3 \ (\textit{Goal State})$$

© The Author(s), under exclusive license to Springer Nature Switzerland AG 2019
M. N. Murty and A. Biswas, *Centrality and Diversity in Search*,
SpringerBriefs in Intelligent Systems, https://doi.org/10.1007/978-3-030-24713-3_3

**Fig. 3.1** Example 3-Puzzle
problem

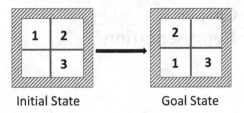

Initial State                              Goal State

This is one way of linear representation of the 2-dimensional view of the puzzle.
There could be other ways of linearizing. To be consistent with the legal moves in
the 3-puzzle, we need to devise equivalent operations on strings.

For example, from the initial state in the figure, it is legally possible to move
the *free space* either *right* (*R*) or *up* (*U*). It is not possible to move the *free space*
either *Left* (*L*) or *Down* (*D*) in this state. The two possible legal operations may be
characterized in the case of strings by

$$1, 2, \#, 3 \to^U \#, 2, 1, 3$$

$$1, 2, \#, 3 \to^R 1, 2, 3, \#$$

The optimal path between the initial state and the goal state using the string representations is

$$1, 2, \#, 3 \to^U \#, 2, 1, 3 \to^R 2, \#, 1, 3$$

With this representation, one can perform all the operations that are possible in
tree/graph representation of the state space.

## 3.3  Vector Space Representation

Representing patterns as vectors is the most popular scheme in machine learning,
pattern recognition, and information retrieval. Here, we will consider how documents
could be represented.

### 3.3.1  What is a Document?

It is important to ask this question as almost anything that you store on a computer
may be viewed as a document as almost any operation one performs on a computer
is some kind of search. A popularly accepted collection of entities that could qualify
as documents includes:

- *Text Documents*:

  1. *Electronic mail (email)*: The quantum of data received through email is increasing enormously. The growth in size is so large to the extent that reading all the maild received during a period of time could become difficult before new mails start accumulating. Both categorization of mail to put them into different folders for easy reference and identifying unwanted/spam mail so that it could be separated from what is typically of interest to the user might require automation.
  2. *Messages*: Typically, text messages are received on mobile devices.
  3. *Multilingual Documents*: Text that employs a combination of languages.
  4. *Tweets*: Twitter is a platform for people to share short messages of text where vocabulary may not be standard.
  5. *Academic publications*: These include books, journal and conference papers, and online course material.

  Different types of text data are represented as vectors.

- *Multimedia Data*

  1. *Web page*: Typically, it includes text, images, video, and an search engines deal with this type of data. The most popular representation is to view them as vectors. Text, images, and video content could be represented as vectors.
  2. *Health Records*: This type of documents might include printed text corresponding to the patient details, handwritten text including doctor's observations and prescriptions, photos, ultrasound images, X-ray, MRI scans, etc.
  3. *Personal Identity Cards*: These may include text, face photo, iris, and fingerprints of the individual.
  4. *Others*: There are several other categories of data, including offline and online news, encyclopedias, product manual pages, company reports, government records, weather reports, and court transcripts.

  All these multimedia data can be represented as vectors for further processing.

- *Software*: This also may be viewed as text data.

  1. *Code*: Typically text in a formal language.
  2. *Documentation*: This will be text in some natural language
  3. *Bug reports*: Typically, multilingual text that combines formal and natural languages.

So, the document is a very pervasive notion and it might be almost any entity/item stored on a computer/network/cloud.

## 3.4 Representing Text Documents

There are different representations employed for text, images, videos, and other data sets. Here, we concentrate on representing text documents. Some of these schemes could be extended to other types of documents. We will examine the representation of network data sets in the applications chapter.

### 3.4.1  Analysis of Text Documents

Earlier researchers have analyzed large collections of documents. The observations are:

1. The frequency, $f_i$, of a term with rank $i$ is given by

$$f_i \propto \frac{1}{i}$$

   where words in the underlying vocabulary are arranged in decreasing order of frequency and the most frequently occurring word like *the* gets rank 1 and the least frequent word gets the rank $|V|$ where $V$ is the vocabulary set. This property is called *Zipf's law*. In information retrieval approaches, it is a common practice to remove the frequent words from the vocabulary. This is called *stopping*.
2. The communication gap between the speaker and the listener is smaller if the listener understands the words used by the speaker.

Noting that frequent words, for example *the, of, to, and*, even though shared well between the speaker and the listener, do not offer much information. On the other hand *rare words* like *biblioklept* if used by the speaker may not be comprehensible to the listener. So, words in the mid-frequency zone are important. This property is exploited by $TF - IDF$ that was discussed earlier which typically will assign larger weights to the words, like *data, mining, pattern, recognition, operating system, etc.* in the mid-frequency zone.

Models that deal with one word at a time are called *uni-word models*. They are popular because of their simplicity resulting out of *bag-of-words* assumption. Here, the order of word occurrence is not important, but the frequency of occurence of a word in the document is important. Sometimes it might be useful to deal with biwords in representing the documents. For example, *machine learning, artificial intelligence, pattern recognition*, and *social networks* are popular biwords. It is possible to consider tri-words (3-word phrases) and beyond. However, this may blowup the number of terms and increase the diversity in representation.

A well-known approach based on $TF - IDF$ weights is to generate soft clusters/features using matrix factorization. Here, the collection of $n$ documents in a $l$-dimensional space ($|V| = l$), is abstracted as a $n \times l$ matrix, $A$, where the $i$th row of $A$ corresponds to the $i$th document, $i = 1, \ldots, n$ and the $j$th column of $A$, is characterized by $j$th term in the vocabulary, $j = 1, \ldots, l$.

The $i, j$th entry in $A$, $A_{ij}$, is the $TF - IDF(i, j)$ value. There are different schemes for factorizing $A$ such that

$$A_{n \times l} = B_{n \times K} C_{K \times l}$$

where $K$ is the number of soft clusters. Additional normalization will also correspond to these soft clusters being viewed as topics. Under this factorization, it is possible to view the $K$ clusters to be bringing out the *Latent Semantics* in the data. There are several schemes for matrix factorization. These include:

- *Latent Semantic Indexing*: Here *singular value decomposition* is used to decompose $A$ as

$$A_{n \times l} = U_{n \times n} D_{n \times l} V_{l \times l}$$

where $U$ and $D$ are obtained based on eigenvectors and eigenvalues of the symmetric matrix $AA^t$ and V is obtained from the eigenvectors of $A^t A$.

- *Random Projections*: The simplest decomposition is

$$A_{n \times l} = B_{n \times K} C_{K \times l}$$

where $B$ is obtained using *Random Projections* of rows of $A$. In other words, $B = A_{n \times l} R_{l \times K}$ where $R$ is a matrix with random entries specified by some distribution. Once $A$ and $B$ are given, obtaining $C$ so that the Frobenius norm, $|A - BC|_F^2$, is minimized is a simpler optimization problem. Using $R$ a pair of documents, $d_i$ and $d_j$, the rows $i$ and $j$ of $A$ are projected to $K$-dimensional vectors, rows $i$ and $j$ of $B$. It is shown that Euclidean distance between a pair of rows of $A$ converges to the Euclidean distance between the corresponding rows of $B$.

- Other Approaches: Some of the other important factorization schemes are based on

    - *Non-negative matrix factorization (NMF)*
      Here matrices $B$ and $C$ are obtained so that they have nonnegative real entries and $|A - BC|_F^2$ is minimized. It is possible because $A$ has nonnegative real entries.
    - *Latent Dirichlet allocation (LDA)*
      Based on some convenient assumptions on the prior distribution, a Bayesian model is used to get $A_{n \times l} = B_{n \times K} C_{K \times l}$. Here, $B$ is a soft assignment matrix indicating how each of the documents is assigned softly to the $K$ topics and $C$ describes the topics. A topic is a probability assignment to words in the vocabulary and each row of $C$ is a topic description in terms of the words in the vocabulary.

Matrix factorization can capture the latent cluster/topic structure in document collections. Based on the number of these clusters/topics (value of $K$), diversity in the representation can get affected; larger $K$ values lead to higher diversity. A major difficulty with the matrix factorization schemes is that they can be computationally explosive and may take an unreasonable amount of time on large data sets.

A more recent and efficient approach is based on representing words as vectors which is called *word2vec* model. Here, a word is converted into a vector based on the occurrence of words in its vicinity in various parts of documents in a document collection. This involves training a neural network with pairs of words where a selected word in the pair is input to the network and the other word is the target output.

Such word pairs are obtained from the corpus by selecting a word and considering words that occur within the vicinity of the selected word for possible pairing. We will provide more details in later chapters.

## 3.5  Representing a Cluster

Typically, hard clustering algorithms are either partitional algorithms or hierarchical. A well-known hard clustering algorithm is the $K$-means algorithm($KMA$). It represents each cluster by its *Centroid*.

### 3.5.1  Centroid

An interesting statistical property of centroid of a cluster of points is that it is closest to all the points in the cluster. Consider a cluster

$$c = \{x_1, x_2, \ldots, x_p\}.$$

Let $x$ be the best vector that minimizes

$$\sum_{i=1}^{p}(x - x_i)^t(x - x_i),$$

that is it minimizes the sum of squared distances to all the $p$ points in the cluster. Equating the gradient to zero would result in

$$x^* = \frac{1}{p}x_i$$

or equivalently $x^*$ is the centroid.

Selection of an appropriate set of centroids is important for the $KMA$ to provide a good partition as discussed earlier. The initial centroids must reflect sufficient diversity and at the same time, they must be representing dense regions; otherwise, outlier clusters will be formed.

Even though centroid summarizes the points in the cluster in an intuitively appealing way, it is not robust to the presence of outliers in the cluster. If one of the $p$ points is an outlier that is far off from the rest of the $p - 1$ points in the cluster, then it can affect the location of the centroid adversely. One solution offered to solve this problem is to consider *medoid* instead of the centroid. Medoid is a point in the cluster that is nearest to all the remaining $p - 1$ points. Medoid is robust to outliers . Note that centroid need not be a member of the cluster whereas medoid is constrained to be a member of the set.

**Fig. 3.2** A dendrogram
depicting hierarchical
clustering

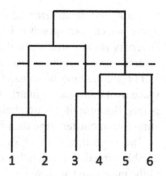

## 3.5.2  Hierarchical Clustering

We depict an abstraction of the hierarchical clustering using an example tree structure, dendrogram as shown in Fig. 3.2.

There are 6 points in this example and they are labeled $1, 2, \ldots, 6$. Each is viewed as a singleton cluster. So, there are 6 clusters as depicted by the leaves. We recursively keep merging a pair of clusters by identifying the most similar pair. Dissimilarity between a pair of clusters $c_i$ and $c_j$ is given by

$$Min_{x \in c_i, y \in c_j} d(x, y).$$

Among all possible pairs of clusters, we select that pair that has the minimum dissimilarity between the two member clusters. Such a merging process is repeated till we are left with a single cluster. This is called the *single-link algorithm*.

In the dendrogram shown in the figure, we depict how starting from 6 singleton clusters, we keep on merging till we are left with a single cluster. Here, the singleton clusters $\{1\}$ and $\{2\}$ are merged to get 5 clusters given by

$$\{\{1, 2\}, \{3\}, \{4\}, \{5\}, \{6\}\}.$$

Among these 5 clusters, $\{3\}$, *and* $\{5\}$ are the most similar and by merging them, we get 4 clusters given by

$$\{\{1, 2\}, \{3, 5\}, \{4\}, \{6\}.$$

Next, the clusters $\{4\}$ and $\{6\}$ are merged followed by merging $\{1, 2\}$ *and* $\{3, 5\}$ are merged to get 2 clusters

$$\{1, 2, 3, 5\}, \{4, 6\}.$$

Next, we merge the two clusters to form a single cluster having all the 6 points. This entire merging process is shown in the figure. Note that each cluster is *represented by a subtree* of the dendrogram in the figure. By stopping the merging process in the dendrogram at an appropriate level, we get the required clusters. It is equivalent to cutting the dendrogram at the required level as shown by the broken line in the figure, which corresponds to having three clusters seen as three subtrees in the figure. Diversity is exhibited when the cut is closer to the leaves and the clustering is less diverse as we shift the cut to be closer to the root.

For a collection of $n$ patterns, hierarchical algorithm requires $\mathcal{O}(n^2)$ time and space as it computes a matrix that stores proximity values between all $\binom{n}{2}$ pairs of points. Because of this quadratic time and space requirements, this algorithm is not attractive to cluster large data sets.

There are a host of clustering algorithms based on different formulations. A notable direction as already mentioned is based on viewing each point as a row of the matrix and achieve clustering through matrix factorization. For example, consider the 3-dimensional data set of 6 points shown as the rows of the $6 \times 3$ matrix $A$ where each of the 3 columns corresponds to one of the 3 features. Consider also its factorization into $B_{6\times2}C_{2\times3}$ given by

$$
A = \begin{bmatrix} 1 & 1 & 1 \\ 1 & 1 & 2 \\ 1 & 2 & 1 \\ 7 & 7 & 6 \\ 7 & 6 & 7 \\ 7 & 7 & 7 \end{bmatrix} \approx \begin{bmatrix} 1 & 0 \\ 1 & 0 \\ 1 & 0 \\ 0 & 1 \\ 0 & 1 \\ 0 & 1 \end{bmatrix} \begin{bmatrix} 1 & 1.33 & 1.33 \\ 7 & 6.66 & 6.66, \end{bmatrix}
$$

where

$$
B = \begin{bmatrix} 1 & 0 \\ 1 & 0 \\ 1 & 0 \\ 0 & 1 \\ 0 & 1 \\ 0 & 1 \end{bmatrix} \quad and \quad C = \begin{bmatrix} 1 & 1.33 & 1.33 \\ 7 & 6.66 & 6.66. \end{bmatrix}
$$

Here, the value of $K$, the number of clusters is 2. Matrix $B$ is the *assignment matrix* depicting the assignment of each of the six patterns to one of the two clusters. Matrix $C$, on the other hand, is the *description matrix* providing the description of each the two clusters (its rows) using the respective centroid.

The data in $A$ is such that the two clusters are reasonably well-separated from each other. So, the hard 2-partition (two clusters) given by

$$
\{(1, 1, 1)^t, (1, 1, 2)^t, (1, 2, 1)^t\}, \{(7, 7, 6)^t, (7, 6, 7)^t, (7, 7, 7)^t\}
$$

is justified. Note that *KMA* would have generated the same 2-partition with the respective centroids, the 2 rows of $C$, summarizing the clusters. So, the above factorization abstracts the *KMA*.

It may not always be the case that hard clustering is justified. Consider another set of five 3-dimensional points shown in matrix $E_{5\times3}$ with a factorization $E = F_{5\times2}G_{2\times3}$ given by

$$E = \begin{bmatrix} 3 & 2 & 3 \\ 2 & 3 & 2 \\ 4 & 4 & 4 \\ 6 & 5 & 6 \\ 5 & 6 & 5 \end{bmatrix} \approx \begin{bmatrix} 1 & 0 \\ 1 & 0 \\ 0.5 & 0.5 \\ 0 & 1 \\ 0 & 1 \end{bmatrix} \begin{bmatrix} 3 & 3 & 3 \\ 5 & 5 & 5. \end{bmatrix}$$

Note that the data in $E$ does not exhibit the same level of well-separated clusters as that in $A$. A possible factorization of $E$ depicting softness in the assignment of the third pattern $(4, 4, 4)^t$ is captured in $F$ and matrix $G$ provides cluster descriptions using some *weighted centroids*.

There are other algorithms that exploit the matrix representations including *spectral clustering*. However, their usage is restricted to smaller data sets. Even explicit matrix factorization algorithms suffer due to computational resource requirements. Currently, there is more interest in *random walk* based methods as they can scale up well. We will discuss them in the application chapter.

## 3.6  Representing Classes and Classifiers

A classifier learning algorithm takes the training data as input and learns an appropriate abstraction. Some of the prominent ones are as follows:

### 3.6.1  Neighborhood Based Classifier (NNC)

In the simplest form of *NNC* all the $n$ training examples in $l$-dimensional space are used to classify a test pattern. So, it is called an *exemplar-based classifier*. It is a simple classifier as it classifies a test pattern $x$ to the class of its *nearest neighbor (NN)* from the training data and it may be represented as $NN(x)$. Equivalently

$$class\_label(x) = class\_label(NN(x)).$$

It is an exemplar-based classifier giving 100% accuracy on the training data. This training data is represented, when there are $C$ classes $c_1, \ldots, c_C$, by

$$\mathcal{X} = \{(x_1, c^1), (x_2, c^2), \ldots, (x_n, c^n)\},$$

where

$$c^j \in \{c_1, \ldots, c_C\} \ for \ j = 1, \ldots, n.$$

Some important properties of *NNC* are as follows:

1. It is often said that it does not employ any *learning* or *knowledge*. However, subtly it does both *learning* and *usage of knowledge*. It has to *abstract or learn* the notion of *neighborhood* of a test pattern from the training data.
2. It exploits the domain knowledge and representation of patterns using an appropriate *distance/similarity or proximity* function.
3. Routinely metrics like the Euclidean distance are used to capture the notion of *neighbor*. Here

$$NN(x) = arg\ min_{x_i \in X}\ d(x, x_i),$$

   where $d$ is a distance function. Typically, $d$ is assumed to be a metric. However, several nonmetric distances like the *KL*-divergence are very popularly used in several machine learning, social network representation, and other areas. What is essential is that different proximity functions capture *neighborhood* in diverse ways; so their usage imposes additional semantics on *neighbor*.
4. A minor practical difficulty with *NNC* is that in order to classify a test pattern $x$ it needs to consider all the $n$ training patterns leading to $\mathcal{O}(n)$ time ans space requirement.
5. A well-known problem is that it is not *robust to outliers* or noise in the training data. Depending on a single *NN* is the problem.
6. The most important difficulty that is not so well-known is that as the dimensionality of the data increases, it fails to identify the correct class label for the test pattern because

$$d(x, NN(x)) \to d(x, FN(x))\ as\ l \to \infty.$$

That is *as the dimensionality increases it is difficult to discriminate between the NN(x) and the farthest neighbor of x (FN(x))*. This limiting behavior is probabilistic. A consequence of this is that in important current day applications, where the dimensionality is large, *NNC* is not a good classifier.

So, *NNC* in its simplest form is associated with some complexity and so is *diversity friendly*. Some variants that can deal with some of these problems are:

1. *CNNC*: It deals with a condensed data set of *prototype patterns* of various classes in classification. This reduces the time and space complexities from $\mathcal{O}(n)$ to $\mathcal{O}(m)$ where $m$ is the number of prototypes in the condensed set. It reduces diversity by possibly selecting the right prototypes. It is not robust to outliers as it adds them to the condensed set.
2. *Minimal distance classifier (MDC)*: The simplest prototype set is obtained by having one representative per class. The representative can be the sample mean/centroid, $\mu$, of the vectors in the class. Now a pattern $x$ is assigned the label of its nearest of the $C$ centroids. This is called the *MDC*. Here,

$$class\_label(x) = class\_label(arg\ min_{\mu_i \in Ptr}\ d(x, \mu_i)),\ where$$

$$Ptr = \{\mu_1, \mu_2, \dots, \mu_C\}.$$

It is possible to show that when the classes are restricted to possess some probability structure, *MDC* will classify *x* exactly like the Optimal classifier.

3. *KNNC*: It is a well-known classifier that classifies a test pattern *x* by finding its label as the majority class label present among *K* nearest neighbors of *x*. So, it uses *K NNs* where typically $K > 1$. So,

    a. It is more robust to noise.

    b. Error rate of *KNNC* can asymptotically ($n \to \infty$) converges to the optimal error rate under conditions on *K*.

    c. It uses *K NNs* of *x* without considering the distances of these neighbors in classifying *x*. To reduce the affect of outliers, it is modified to get weighted majorities of different class labels among the *K NNs* by weighing the contributions of the neighbors based on their neighborhood ranks. The nearer the neighbor of *x* among the *K* neighbors the larger its importance (weight).

    d. In practice, *n* is finite and if $K = n$ then any test pattern *x* is labeled with the class label of the majority class in the training data. This is a very simple model and exhibits centrality. As the value of $K \to 1$, it offers more diversity. Smaller the value of *K* larger the diversity.

4. *Fractional Norms*: It is observed that instead of using Euclidean distance to get the *NN* (*x*), city block distance ($L_1$ norm) will do better as the dimensionality grows. Further, fractional norms can do better than the $L_1$ norm on high-dimensional data. Here, the fractional norm, *d*, is defined as

$$d(x, x_i) = (\sum_{j=1}^{l} |x(j) - x_i(j)|^q)^{\frac{1}{q}},$$

where *q* is a fraction in (0, 1) and *x*(*j*) is the *j*th component of *x*, for $j = 1, \ldots, l$. So, fractional norms reduce the diversity in classification by *NNC* and its variants.

5. *Approximate NN*: There are several schemes suggested to get an approximate *NN* of a test pattern *x*. One recent contribution in this direction is the *locality sensitive hashing (LSH)* which could be viewed simply as obtaining the *NN* in a randomly selected subset of features. It combines the approximate *NNs* obtained by some *L* random subsets of features. So, neighborhood is defined here based on a collection of neighbors in several random subspaces. This combination can reduce the diversity in *NNC* as it operates in lower dimensional subspaces.

*NNC* captures all the details including noise and so is more diverse compared to either *CNNC* or $K - NNC$. In the case of *CNNC*, only essential patterns from the training data are retained; so it is a simpler model. In the case of *KNNC*, the value of *K* is learnt using the training and validation data. Along with increase in the value of *K*, performance of *KNNC* becomes more predictable. Even though NNC is a simple classifier, it has an error rate that is twice that of the optimal Bayes classifier asymptotically. So, in classification of big data, NNC is more robust. However, it is computationally inefficient as its abstraction capability is poor.

### 3.6.2  Bayes Classifier

Here, each class is represented by a probability density function. With the availability of the prior probabilities it computes the posterior probability using Bayes rule and uses the posterior probabilities to come out with an optimal classifier. If $P(c_i)$ is the prior probability of class $c_i$ and $P(x/c_i)$ is the mass function of $x$ under class $c_i$, then the posterior probability $P(c_i/x)$, that depends upon a specific $x$ to be classified is

$$P(c_i/x) = \frac{P(c_i)P(x/c_i)}{\sum_{j=1}^{C} P(c_j)P(x/c_j)}.$$

It computes the posterior probability for each of the $C$ classes and assigns $x$ to the class with the largest posterior probability. It minimizes the *average of the probability of error*. However, it needs to estimate the underlying probability structure from the training data. Naïve Bayes classifier (*NBC*) is a simplified model where features are assumed to be class-conditionally independent. That is if $x$ is an $l$-dimensional vector then $P(x/c_i)$ is approximated as $\prod_{p=1}^{l} P(x_p/c_i)$. This assumption simplifies the estimation of the probabilities. Bayes classifier is based on the probability structure and is an ideal one. However, to realize it, we need to depend on statistics and so is affected by the empirical properties including the *bias-variance* trade-off.

### 3.6.3  Neural Net Classifiers

There are a variety of artificial neural net classifiers including perceptron, multilayer perceptron (MLP), and support vector machine (SVM).

#### 3.6.3.1  Perceptron

It is the simplest neural classifier. In its simplest from, it may be viewed as a *linear classifier*.

If $x$ is a $l$-dimensional point, then perceptron finds a $l$-dimensional vector $w$ and a scalar $b$ such that

$$w^t x + b > 0 \; if \; x \in c_+ \; and$$

$$w^t x + b < 0 \; if \; x \in c_-.$$

The second inequality can be rewritten as

$$-w^t x - b > 0 \; if \; x \in c_-.$$

So, both these inequalities can be replaced by a single inequality of the form

$$w_a^t x_a > 0$$

where $w_a = [w, b]$ and $x_a = [x, 1]$ if $x \in c_+$ and $x_a = [-x, -1]$ if $x \in c_-$ where $w_a$ and $x_a$ are suitably appended $l + 1$-dimensional vectors. Even though $w_a$ and $w$ are different because $w^t x + b = w_a^t x_a$, we will use $w$ for the augmented vector $w_a$ also and the usage will be clear based on the context in which it is used. Some details of the perceptron algorithm (Algorithm 4) for obtaining a $w$ from the training data will be discussed next.

- *Perceptron Learning Algorithm (PLA)*

---

**Algorithm 4** Perceptron Learning

---

**Input:** A labeled and augmented training data set $\mathcal{X} = \{(x_1, lab_1), (x_2, lab_2), \ldots, (x_n, lab_n)\}$ with $n$ data objects; $lab_i$ is the *class label* of $x_i$. There are two classes—a positive class and a negative class.
**Output:** A vector $w$ such that $w^t x_i > 0$, $\forall i$.
1: Let $k = 0$ and $w_k = 0$. Consider patterns in the order $x_1, \ldots, x_n, x_1, \ldots, x_n, x_1, \ldots$.
2: STOP if no pattern in a sequence of $n$ successive patterns is misclassified by $w_k$; declare $w = w_k$.
   Else let $x_p$ be the first misclassified pattern in the order by $w_k$ or equivalently $w_k^t x_p \leq 0$.
3: Let $k = k + 1$, and $w_k = w_{k-1} + x_p$. Goto step (2).

---

- It is possible to show that if there is at least one such $w$ that exists satisfying the inequality for all the training patterns in a two class ($c_+$, *and* $c_-$) problem, then perceptron can learn such a $w$. Such a pair of classes are called *linearly separable*.
- *PLA* is a deterministic algorithm and it converges to a $w$ in a finite number of iterations (*epochs*) over the data if the two classes are linearly separable.
- **Example 3**
  Let us consider a 1-dimensional labeled data set

$$\{(1, c_-), (2.1, c_-), (3.5, c_-), (5.7, c_+), (6.5, c_+).$$

In this example,

$$x \leq 3.5 \Rightarrow x \in c_- \quad and \quad x \geq 5.7 \Rightarrow x \in c_+$$

Based on the training data, a simple rule like $x > 4 \Rightarrow x \in c_+$ else $x \in c_-$ is adequate. Instead of 4, any value in the interval $(3.5, 5.7)$ will work. The augmented data is

$$(-1, -1)^t, (-2.1, -1)^t, (-3.5, -1)^t, (5.7, 1)^t, (6.5, 1)^t.$$

Starting with $w_0 = (0, 0)^t$, in a small number of updates we get $w = (3.7, -13)^t$ which means $(3.7, -13).(x, 1) = 3.7x - 13 = 0$ defines the boundary between the two classes. Equivalently, $x = \frac{130}{37} \approx 3.513$. So, a value less than 3.513 belongs to

$c_-$, else it belongs to $c_+$. If one goes through the updates on $w$ using *PLA* it will be noticed that the number of updates is larger because of $(-3.5, -1)^t$ that is closest to the boundary, 3.513, between $c_-$ and $c_+$. The reason is that $p = (-3.5, -1)^t$ being closest to the decision boundary can be close to being orthogonal to the weight vector $w$ and hence $w^t p$ is close to 0; so, it can have more impact on the convergence of *PLA*.

### 3.6.3.2  Multilayer Perceptron (*MLP*)

Perceptron fails to capture the discrimination information between the classes when they are not linearly separable. For example, consider the boolean function shown in Table 3.1. Note that the boolean function $f(x, y)$ and the algebraic form $2xy - x - y + 1$ are equivalent as shown in the table. Further, the $xy$ term in the expression in column 4 of the table indicates the possible nonlinearity.

The input–output behavior of the function is shown in Table 3.1. Note that it is not possible to linearly separate the two points with output 1 from the other two points with output 0 or equivalently there is no straight line segment that can separate the 0s from 1s. However, from the equivalent function, $2xy - x - y + 1$ shown in the table, it is possible to have a linear separation in the 3-dimensional space of $(xy, x, y)$. Specifically, if $z = (xy, x, y, 1)^t$ and $W_m = (2, -1, -1, 1)^t$, then

$$f(x, y) = W_m^t z$$

is a linear function in the 4-dimensional space.

This view perhaps hints at using the perceptron with four inputs $xy, x, y$; *and* 1 to do the job. However, computing $xy$ from $x$ and $y$ needs one more perceptron unit as shown in Fig. 3.3. This leads to a two-layer perceptron network which is an *MLP*. Typically, such an *MLP* is trained using backpropagation algorithm which is a gradient descent-based algorithm that updates vector $W$ to minimize the error between the target output and output obtained using the current $W$ vector. This gradient descent approach may not get the best $W$; the backpropagation algorithm may reach a *local minimum* of the error. The diversity in descent direction is controlled by using some heuristics like *momentum term*.

**Table 3.1**  Typical patterns

| $x$ | $y$ | $f(x, y)$ | $2xy - x - y + 1$ |
|---|---|---|---|
| 0 | 0 | 1 | 1 |
| 1 | 0 | 0 | 0 |
| 0 | 1 | 0 | 0 |
| 1 | 1 | 1 | 1 |

**Fig. 3.3** An MLP for the
boolean function

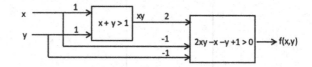

### 3.6.3.3 Support Vector Machine

An MLP once trained may get the best $W$ vector or equivalently train the network properly. This motivated the need for a better learning algorithm in the form of the *support vector machine (SVM)*. A more convenient criterion function is chosen for optmization by SVM.

Here, the binary classification problem is formulated as maximization of the *margin* or separation between the two classes. In case there is no margin between the two classes because there is an overlap between the classes, a *soft margin formulation* is used to reduce the impact of the points on the margin, or the non-feasible points. A consequence of this is that the resulting optimization function is well behaved. The weight vector $W$ learnt by $SVM$ is based on typically a small set of boundary patterns and margin violators which are called *support vectors*. These support vectors are demonstrated to form some kind of essential set or support set for classification, clustering, and regression.

### 3.6.4 Decision Tree Classifiers (DTC)

Decision tree classifier may be viewed as a piece-wise linear classifier that combines a sequence of decisions to abstract the training data into the tree. The entire training data set is associated with the root node of the tree. At each nonterminal node, including the root node of the tree, a test based on the best feature is used to split the training data set associated with the node into smaller data sets and associate them with its children nodes. Here, the best feature and the split are based on minimizing entropy in the resulting split. The split is axis parallel, that is using one feature at a time. For example, consider the data set shown in Fig. 3.4.

The corresponding decision tree is shown in Fig. 3.5. All the 11 patterns (6 x, 5 o) are associated with the root nodes in both (a) and (b) in the figure. The axis-parallel split on feature $f < a$ in Fig. 3.5a is better than the axis-parallel split using $f'$ shown in Fig. 3.5b. Note that against each node of the trees, the number of patterns in each class is displayed to compute the proportions of the two classes. This may be explained using *gini index*, an impurity measure associated with a node.

Gini index is based on the proportion of patterns, from each class, at a node. If there are $C$ classes in the data and the proportions of points at a node $v$ are $p_1, p_2, \ldots, p_C$, then gini index (GI) at $v$ is

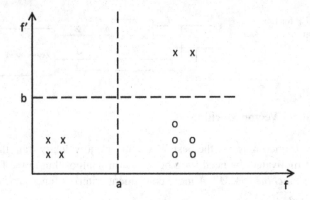

**Fig. 3.4** Example data set

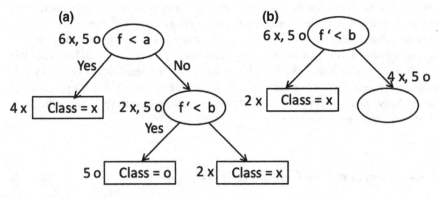

**Fig. 3.5  a** The decision tree constructed **b** Another split at the root

$$GI(v) = 1 - \sum_{i=1}^{C} p_i^2.$$

For example, in Fig. 3.5, there are two classes ($C = 2$). Out of the 11 patterns associated with the root in (a), the left child has 4 points all from class x. So the proportions are $p_x = 1$ and $p_o = 0$. So, its GI value is 0 indicating that the node is pure. In the case of the right child of the root in (a) in the figure, the proportions are $p_x = \frac{2}{7}$ and $p_o = \frac{5}{7}$. So, the GI value of the node is $\frac{20}{49}$. Hence, the split based on $f$ at the root of the tree in (a) of the figure leads to weighted impurity of

$$0.\frac{4}{11} + \frac{20}{49} \cdot \frac{7}{11} = \frac{20}{77}$$

as the left child has 0 impurity and the right child with 7 out of 11 points has impurity of $\frac{20}{49}$.

On the contrary, in (b) of the figure, where ta the root the split is based on $f'$, the split has a weighted impurity of

$$0.\frac{2}{11} + \frac{9}{11}.(1 - (\frac{4}{9})^2 - (\frac{5}{9})^2) = \frac{40}{99}.$$

So, the impurity is more than that corresponding to the split at the root in (a) of the figure.

In general, the root node carries the decision based on the *best feature* that leads to the least impurity split. The construction of the decision tree is done in a greedy manner. So, once the best decision is fixed for the root, the splits at the nodes that are descendants are biased by the decision made at the root node. The implications of this greedy decision in building the tree are:

1. If the classification problem is high-dimensional, then to arrive at the best decision, all the features in the large set need to be considered for evaluating the best split. This limits the use of decision tree classifiers to low-dimensional problems.
2. Even in the low-dimensional problems, the greediness may lead to decisions with many levels to ensure purity at the leaf level. It is demonstrated that trees with a single decision node perform well on a large number of benchmark data sets.

Some of the important positive aspects of decision tree classifier are as follows:

1. It is highly transparent classifiers and are simple.
2. It can be extended to be used for both classification and regression like the *KNNC*.

Even though it is possible to use splits other than axis-parallel splits, the alternatives are not popular because of computational complexity. Also, decision trees with too many levels do not do well in classifying the test patterns (generalization) as the trees may overfit the training data. Such classifiers exhibit a lot of diversity; minor changes in the training data may lead to variations in the trees generated. This diversity is controlled by pruning the leaf nodes in the tree which account for a small number of points. Equivalently splitting at nodes that have a very small amount of impurity is not performed. This improves the generalization by making the tree classifier simpler.

Simpler decision tree classifiers are combined to learn more robust classifiers as follows:

1. Several simple decision tree classifiers are combined to form a classifier that performs well. In fact, it is possible to combine decision stump classifiers (one node decision trees) and realize a competitive classifier.
2. To deal with high-dimensional data sets having $n$ patterns and $l$ features, a forest of simple decision trees is used in classification. Each decision tree classifier is built from $n$ training patterns of $l'$-dimension with $l' \ll l$ with possible repetitions (sampling with replacement). A collection of $p$ such decision tree classifiers are used to classify each test pattern by using the majority class label obtained from the $p$ classifiers on the test pattern. Such a forest of classifiers is popularly called as the *random forest (RF) classifier* which controls the diversity in classification.

## 3.7   Summary

In this chapter, the notion of representation of problem states, patterns, clusters, and classes is considered. This has an impact on the underlying tasks in AI, machine learning, and information retrieval. The roles of centrality and diversity in searching for proper representations is also discussed. We summarize some of these roles in Table 3.2.

Recall the discussion we had on the role of *bias and variance* in learning classifiers. In a nutshell, simpler models lead to *low variance and central* models and complex models are *low bias and diverse* models. We summarize how the representation of data, and classes can impact the classifier's behavior. In Table 3.3, we consider *NNC* and its variants.

In the case of classifiers other than *NN* based also, representation can impact the centrality and diversity in searching for representation of patterns, classes, and other hyperparameters as shown in Table 3.4.

**Table 3.2**  Centrality and diversity in representation

| Task | Centrality | Diversity |
|------|-----------|-----------|
| AI problem-solving | Tree (simpler search) | Graph (complex search) |
| Document processing | Mid-frequency words (low-dimensional) | Stopping (Zipf's law) (high-dimensional) |
| Phrase length | Bag-of-words | N-words |
| Document matrix factorization | Smaller number of topics | Large number of topics |
| K-means clustering | Medoid based (robust to outliers) | Centroid based (outliers affect) |
| Single-link clustering (Hierarchical) | Dendrogram is cut close to root | Dendrogram is cut close to leaves |

**Table 3.3**  Centrality and diversity in neighborhood based classifiers

| Classifier | Diversity | Centrality |
|-----------|-----------|-----------|
| NNC | Diverse (noise, outliers) | Simpler (approximations) |
| CNNC | More diverse (outliers) | Less diverse (Apt prototype set) |
| KNNC | Larger diversity ($K$ small) | Predictable ($K$ is large) |
| Distance | Larger (Euclidean and higher) | Simpler $L_1$ and fractional |
| LSH | Less diverse | Simpler (central) |

**Table 3.4** Centrality and diversity in classification

| Classifier | Diversity | Centrality |
|---|---|---|
| *Bayes classifier* | Optimal (class-conditional pdf) | Optimal (Prior Probabilities) |
| *NBC* (small data sets) | Larger (max likelihood estimation) | Simpler (Bayesian estimation) |
| *NBC* (Big data) | Smaller (max likelihood estimation) | Simpler (Bayesian estimation) |
| MLP | Higher (local minimum of error) | Higher (regularizer like *momentum*) |
| Perceptron | Larger ($W$ varies) | Smaller |
| *SVM* | Smaller (smaller variation in $W$) | Larger |
| *DTC* | Larger (taller tree) | Larger (pruned tree) |
| *RF* (low variance) | Smaller | Larger |

# Bibliography

1. Zipf GK (1949) Human behavior and the principle of least effort. Addison-Wesley
2. Murty MN, Devi VS (2015) Introduction to Pattern recognition and machine learning. IISc Press
3. Fukunaga K (2013) Introduction to statistical pattern recognition. Academic Press
4. Duda RO, Hart PE, Stork DG (2001) Pattern classification. Wiley Interscience
5. Francois D, Wertz V, Verleysen M (2007) The concentration of fractional distances. IEEE Trans Knowl Data Eng 19(7):873–886
6. Andoni A, Indyk P (2008) Near-optimal hashing algorithms for approximate nearest neighbor in high dimensions. Commun ACM 51(1):117–122
7. Murty MN, Raghava R (2016) Support vector machines and perceptrons. Springer briefs in computer science, Springer, Cham
8. Holte RC (1993) Very simple classification rules perform well on most commonly used datasets. Mach Learn 11:63–91

# Chapter 4
# Clustering and Classification

**Abstract** Optimization is another important tool that helps in defining, designing, and in model selection in various machine learning tasks including dimensionality reduction, clustering, and classification. We discuss, in this chapter, the role of optimization in feature selection, feature extraction, clustering, and classification.

**Keywords** Optimization · Regularization · Feature selection · Classification · Clustering

## 4.1 Introduction

We have examined the roles of centrality and diversity in search and representation earlier. In the discussion, we had considered their roles in clustering and classification also. In this chapter, we will consider more details on the roles of centrality and diversity in clustering and classification.

## 4.2 Clustering

We have observed that a *unifying representation* of both hard and soft clustering is through matrix factorization. Specifically, if we represent the set of $n$ $l$-dimensional points,

$$\mathcal{X} = \{x_1, x_2, \ldots, x_n\},$$

to be clustered as the rows of a matrix $A_{n \times l}$, then we can factorize it into the product of matrices $B_{n \times K}$ and $C_{K \times l}$ where

- $B_{n \times K}$ is the *cluster/topic assignment matrix*, $B_{ik}$ is the membership or importance of cluster $k$ to pattern $i$ for $i = 1, \ldots, n$ and $k = 1, \ldots, K$.

- $C_{K \times l}$ is the *cluster/topic description matrix* where $C_{kj}$ indicates the importance of feature $j$ to cluster $k$, $j = 1, \ldots, l$ and $k = 1, \ldots, K$.

An observation in such a representation is that any *data matrix* $A_{n \times l}$ is a $\mathfrak{R}^{n \times l}$ structure. Further, any $n \times l$ matrix $A$ has its

$$rank(A) = row - rank(A) = column - rank(A)$$

where row rank is the number of *linearly independent* rows in $A$ and column rank is the number of linearly independent columns of $A$. Because clustering is grouping of rows ($n$ patterns) and dimensionality reduction deals with columns ($l$ features), their ranks being equal means the number of clusters and number of features are equal from the *linear independence* view.

We examine one representative each from clustering, feature selection, and feature extraction with the help of an example data set.

*Example 4.1* Let $A_{4 \times 3}$ be a matrix representing 4 patterns in a 3-dimensional space given by

$$A = \begin{bmatrix} 1 & 0 & 1 \\ 0 & 1 & 0 \\ 1 & 0 & 1 \\ 0 & 1 & 0 \end{bmatrix}$$

For the sake of simplicity rows are replicated, rows 1 and 3 are identical and rows 2 and 4 are also same. We will examine clustering first using matrix $A$.

### 4.2.1   Clustering-Based Matrix Factorization

Clustering the 4 rows of $A$ into $K = 2$ clusters gives us $\{A_1, A_3\}$ and $\{A_2, A_4\}$, where $A_i$ is the $i$th row of $A$. This is obtained by selecting the first two rows, diverse rows, as the initial cluster centers and assigning the remaining two points, third and fourth rows based on nearness to the selected points. The centroids of clusters $c_1$ and $c_2$ are $(1, 0, 1)$ and $(0, 1, 0)$ respectively. This gives us

- The assignment matrix $B_{4 \times 2}$ to be

$$B = \begin{bmatrix} 1 & 0 \\ 0 & 1 \\ 1 & 0 \\ 0 & 1 \end{bmatrix}$$

- The cluster description matrix $C_{2 \times 3}$ has the 2 centroids as its rows given by

$$C = \begin{bmatrix} 1 & 0 & 1 \\ 0 & 1 & 0 \end{bmatrix}$$

- Note that

$$A = \begin{bmatrix} 1 & 0 & 1 \\ 0 & 1 & 0 \\ 1 & 0 & 1 \\ 0 & 1 & 0 \end{bmatrix} = \begin{bmatrix} 1 & 0 \\ 0 & 1 \\ 1 & 0 \\ 0 & 1 \end{bmatrix} \begin{bmatrix} 1 & 0 & 1 \\ 0 & 1 & 0 \end{bmatrix} = BC$$

Observe that because of the simplicity of the data, any clustering algorithm will lead to the same partition and if centroids of clusters or other representatives are used, again we get the same $C$ matrix. However, some important observations are

- In general $A \approx BC$. In this example, $A = BC$ because each centroid coincides with 2 out of the 4 patterns.
- In most of the practical applications $A$ will have elements from $\Re^+ \cup \{0\}$. The factorization is called non-negative matrix factorization (*NMF*) if elements of $B$ and $C$ are nonnegative reals.
- It is known that in such a *NMF* if any two out of $A$, $B$, $C$ are given, then getting the third one is simple. In *KMA* based clustering, given $A$, getting the centroids and the $C$ matrix are reasonably straightforward.
- In *NMF*, in general, we are given $A$ and finding $B$ and $C$ is posed as the optimization problem

$$\min_{B,C} ||A - BC||_F \ s.t. B \geq 0, C \geq 0$$

where $||A - BC||_F$ is the squared Frobenius norm or element-wise difference between the $n \times l$ matrices $A$ and $BC$.

### 4.2.2 Feature Selection

It is easy to observe that columns 1 and 3 are identical in matrix $A$. So, by grouping the columns and identifying diverse columns gives rise to using either columns 1 and 2 or columns 2 and 3. Suppose we use columns 1 and 2 to represent matrix $B$, then

$$B = \begin{bmatrix} 1 & 0 \\ 0 & 1 \\ 1 & 0 \\ 0 & 1 \end{bmatrix}.$$

Consequently we get the same $C$ matrix as earlier that is given by

$$C = \begin{bmatrix} 1 & 0 & 1 \\ 0 & 1 & 0 \end{bmatrix}.$$

This simple example is ideally suited to illustrate the equivalence between features and clusters by using feature selection. Further, all the matrices involved are non-negative. So, this is an example *NMF*.

### 4.2.3   Principal Component Analysis (PCA)

Principal components, *PCs*, are popular linear feature extractors. Given the data represented in $l$-dimensional space using features $f_1, f_2, \ldots, f_l$. An extracted feature, $f$, is a linear combination that is obtained from the given $l$ features. So, $f = \sum_{i=1}^{l} \alpha_i f_i$ where $\alpha_i$ is the weight or importance associated with the given feature $f_i$. In general, we can extract features using nonlinear combinations also, but that may be time consuming.

In *PCA*, the features extracted are the eigenvectors of the covariance matrix of the data. These are popularly called the principal components (*PCs*). There could be up to $l$ PCs when $A$ is an $n \times l$ matrix. These are ordered based on decreasing order of the respective eigenvalues. Some properties of *PCA* are

1. Because the underlying matrix is the covariance matrix, these eigenvalues are variances in the direction of the respective *PCs*. So, the first *PC* is in the maximum variance direction of the data.
2. The covariance matrix is a *symmetric matrix*. So, the eigenvectors (*PCs*) are orthogonal to each other when the corresponding eigenvalues are distinct.
3. If we take the first $K$ out of $l$ possible *PCs* to represent the data, it corresponds to optimizing a criterion function that captures average deviations between the given patterns in the $l$-dimensional space and a $K$-dimensional space. This minimization leads to $K$ *PCs* as the optimal new features that are linear combinations of the given features.
4. These *PCs* provide uncorrelated directions under some conditions.

Considering the data matrix $A_{4\times3}$, the corresponding sample covariance matrix is obtained first by getting the zero-mean normalized matrix, $A^n$ is

$$
A^n = \begin{bmatrix}
\frac{1}{2} & -\frac{1}{2} & \frac{1}{2} \\
-\frac{1}{2} & \frac{1}{2} & -\frac{1}{2} \\
\frac{1}{2} & -\frac{1}{2} & \frac{1}{2} \\
-\frac{1}{2} & \frac{1}{2} & -\frac{1}{2}
\end{bmatrix}
$$

and then the covariance matrix $\Sigma$ given by $A^{nt}A^n$ which is

$$
\Sigma = \frac{1}{4} \begin{bmatrix}
1 & -1 & 1 \\
-1 & 1 & -1 \\
1 & -1 & 1
\end{bmatrix}.
$$

The eigenvalues of $\Sigma$ are 3, 0, and 0. So, the top two eigenvectors are $(1, -1, 1)^t$ and $(1, 2, 1)^t$. They are orthogonal. To make them orthonormal we normalize them to make them *unit norm* vectors to get the two *PCs* to be

$$\left(\frac{1}{\sqrt{3}}, -\frac{1}{\sqrt{3}}, \frac{1}{\sqrt{3}}\right)^t, \left(\frac{1}{\sqrt{6}}, \frac{2}{\sqrt{6}}, \frac{1}{\sqrt{6}}\right)^t$$

So, $C_{pc}$ matrix is given by

$$C_{pc} = \begin{bmatrix} \frac{1}{\sqrt{3}} & -\frac{1}{\sqrt{3}} & \frac{1}{\sqrt{3}} \\ \frac{1}{\sqrt{6}} & \frac{2}{\sqrt{6}} & \frac{1}{\sqrt{6}} \end{bmatrix}.$$

This gives us the $B_{pc}$ matrix to be

$$B_{pc} = \begin{bmatrix} \frac{2}{\sqrt{3}} & \frac{2}{\sqrt{6}} \\ -\frac{1}{\sqrt{3}} & \frac{2}{\sqrt{6}} \\ \frac{2}{\sqrt{3}} & \frac{2}{\sqrt{6}} \\ -\frac{1}{\sqrt{3}} & \frac{2}{\sqrt{6}} \end{bmatrix}.$$

Note that the 4 rows of $B_{pc}$ are obtained by projecting the 4 patterns onto these two *PCs*. Projecting the first row (pattern) of $A$, that is $(1, 0, 1)$ gives us $(\frac{2}{\sqrt{3}}, \frac{2}{\sqrt{6}})$. The second row projection gives us $(-\frac{1}{\sqrt{3}}, \frac{2}{\sqrt{6}})$. Putting them all together, we have $A = B_{pc}C_{pc}$ given by

$$A = \begin{bmatrix} 1 & 0 & 1 \\ 0 & 1 & 0 \\ 1 & 0 & 1 \\ 0 & 1 & 0 \end{bmatrix} = \begin{bmatrix} \frac{2}{\sqrt{3}} & \frac{2}{\sqrt{6}} \\ -\frac{1}{\sqrt{3}} & \frac{2}{\sqrt{6}} \\ \frac{2}{\sqrt{3}} & \frac{2}{\sqrt{6}} \\ -\frac{1}{\sqrt{3}} & \frac{2}{\sqrt{6}} \end{bmatrix} \begin{bmatrix} \frac{1}{\sqrt{3}} & -\frac{1}{\sqrt{3}} & \frac{1}{\sqrt{3}} \\ \frac{1}{\sqrt{6}} & \frac{2}{\sqrt{6}} & \frac{1}{\sqrt{6}} \end{bmatrix}.$$

This factorization is indicating how the 3-dimensional points are represented in the 2-dimensional *PC* space. When the rank of the matrix $A$ is 2, which is the case here, we can represent it using 2 orthogonal basis vectors as indicated in the equality between $A$ and $B_{pc}C_{pc}$. Also this is not an *NMF* as there are negative elements in both $B_{pc}$ and $C_{pc}$.

However, the second eigenvalue of $\Sigma$ is 0. So, the variance is captured by the first *PC* itself. In such a case, using the first *PC* we get approximation

$$A = \begin{bmatrix} 1 & 0 & 1 \\ 0 & 1 & 0 \\ 1 & 0 & 1 \\ 0 & 1 & 0 \end{bmatrix} \approx \begin{bmatrix} \frac{2}{\sqrt{3}} \\ -\frac{1}{\sqrt{3}} \\ \frac{2}{\sqrt{3}} \\ -\frac{1}{\sqrt{3}} \end{bmatrix} \begin{bmatrix} \frac{1}{\sqrt{3}} & -\frac{1}{\sqrt{3}} & \frac{1}{\sqrt{3}} \end{bmatrix}.$$

Here $B_{pc}$ is the projection of the 4 rows of $A$ onto the first $PC$.

This approximation amounts to $||A - B_{pc}C_{pc}||_F = \frac{16}{3}$, where each pattern is approximated with an error of $\frac{4}{3}$. However, the 1-dimensional representation is able to discriminate between the patterns 1 and 3 from the patterns 2 and 4. There could be other approximations with a lesser value of 4 as the squared Frobenius norm for the following.

$$A = \begin{bmatrix} 1 & 0 & 1 \\ 0 & 1 & 0 \\ 1 & 0 & 1 \\ 0 & 1 & 0 \end{bmatrix} \approx \begin{bmatrix} \sqrt{3} \\ 0 \\ \sqrt{3} \\ 0 \end{bmatrix} \begin{bmatrix} \frac{1}{\sqrt{3}} & -\frac{1}{\sqrt{3}} & \frac{1}{\sqrt{3}} \end{bmatrix} = \begin{bmatrix} 1 & -1 & 1 \\ 0 & 0 & 0 \\ 1 & -1 & 1 \\ 0 & 0 & 0 \end{bmatrix}.$$

Even though some discrimination between elements of the two clusters is exhibited in the $PCs$ space, in general the first $K$ $PCs$ may not be able to retain the discrimination present in the $l$-dimensional space. The reason is that the underlying optimization is planned to reduce the expected squared deviation between the patterns in the $l$-dimensional space and their representations in the $K$-dimensional space specified by minimization of

$$E[(x^l - x^K)^t(x^l - x^K)],$$

where $x^l$ and $x^K$ are original pattern and its approximation, that is represented in the $K(< l)$ dimensional space respectively and $E$ is the expectation operation. The following high-level summary of the properties will link the above criterion function and the $PCs$.

- Note that $x^l$ is a vector in a $l$-dimensional space. So, it can be uniquely represented using $l$ orthonormal basis vectors $v_1, \ldots, v_l$. Specifically,

$$x^l = \sum_{i=1}^{l} d_i v_i$$

where $d_i$s are some real numbers, for $i = 1, \ldots, l$.
- Now $x^K$ may be viewed as coming out of $K$-dimensional subspace and

$$x^K = \sum_{i=1}^{K} d_i v_i$$

- The error, by exploiting the orthonormality property of $v_1, v_2, \ldots, v_l$ will reduce to

$$error = E[(x^l - x^K)^t(x^l - x^K)] = \sum_{i=K+1}^{l} v_i^t \Sigma v_i = \sum_{i=K+1}^{l} v_i^t v_i \lambda_i = \sum_{i=K+1}^{l} \lambda_i$$

where $v_i$ and $\lambda_i$ are an eigenvector and the respective eigenvalue of $\Sigma$.

- This error is minimized when $\lambda_{K+1}, \lambda_{K+2}, \ldots, \lambda_l$ are smaller. This indicates that $\lambda_1, \lambda_2, \ldots, \lambda_K$ need to be the larger eigenvalues. Correspondingly, $v_1, v_2, \ldots, v_K$ are the eigenvectors that characterize $x^K$.
- So, first $K$ PCs are the eigenvectors of $\Sigma$ which can uniquely characterize projection of each pattern in the $K$ space.

So, error considered is intuitively appealing as it minimizes the average error between patterns in the $l$ space and the respective projections in the $K$ PCs space. This optimization is reproduction friendly and the basis vectors in the $K$ space capture the variance in the data to the best possible extent. However, there is no guarantee that the $K$ PCs retain the discrimination present in the patterns.

### 4.2.4  Singular Value Decomposition (SVD)

A more general factorization of $A_{n \times l}$ may be viewed $A_{n \times l} = B_{n \times n} D_{n \times l} C_{l \times l}$, where $D$ is a diagonal matrix with $n - l$ zero rows if $n > l$ or with $l - n$ zero columns if $n < l$. In the earlier cases, where $A = BC$, $D$ may be viewed as having in its diagonal portion the identity matrix $I$.

SVD may be viewed as

- orthonormal eigenvectors of the symmetric matrix $AA^t$ as the columns of $B$.
- orthonormal eigenvetors of the symmetric matrix $A^tA$ as the rows of $C$.
- Square roots of the eigenvalues of $AA^t$ or $A^tA$, based on whether $n < l$ or $l < n$ respectively, as the diagonal entries of $D$ with remaining elements to be 0. These diagonal entries are called the singular values of $A$.
- Importantly, $SVD$ always gives $B$, $D$, *and* $C$ such that $A = BDC$, an exact deterministic factorization of any $A$ matrix.

Consider the matrix $A$ given in the example, we have

$$A^tA = \begin{bmatrix} 2 & 0 & 2 \\ 0 & 2 & 0 \\ 2 & 0 & 2 \end{bmatrix}.$$

The eigenvalues of $A^tA$ are 4, 2, *and* 0 the respective eigenvectors are $(1, 0, 1)^t$, $(0, 1, 0)^t$, $(1, 0, -1)^t$. They are orthogonal and by normalizing them to be unit norm vectors, we get the $C$ matrix as

$$C = \begin{bmatrix} \frac{1}{\sqrt{2}} & 0 & \frac{1}{\sqrt{2}} \\ 0 & 1 & 0 \\ \frac{1}{\sqrt{2}} & 0 & -\frac{1}{\sqrt{2}} \end{bmatrix}$$

Similarly, the eigenvalues of $AA^t$ are 4, 2, 0, 0 and respective orthonormal eigenvectors that are used as columns of $B$ give $B$ as

$$B = \begin{bmatrix} \frac{1}{\sqrt{2}} & 0 & \frac{1}{2} & -\frac{1}{2} \\ 0 & \frac{1}{\sqrt{2}} & -\frac{1}{2} & -\frac{1}{2} \\ \frac{1}{\sqrt{2}} & 0 & -\frac{1}{2} & \frac{1}{2} \\ 0 & \frac{1}{\sqrt{2}} & \frac{1}{2} & \frac{1}{2} \end{bmatrix}.$$

The $D_{4 \times 3}$ is given

$$D = \begin{bmatrix} 2 & 0 & 0 \\ 0 & \sqrt{2} & 0 \\ 0 & 0 & 0 \\ 0 & 0 & 0 \end{bmatrix},$$

where nonzero entries $\sqrt{4} = 2$, *and* $\sqrt{2}$ are the *singular values* that are the positive square roots of the nonzero eigenvalues of either $AA^t$ or $A^tA$. Note that

$$A = \begin{bmatrix} 1 & 0 & 1 \\ 0 & 1 & 0 \\ 1 & 0 & 1 \\ 0 & 1 & 0 \end{bmatrix} = \begin{bmatrix} \frac{1}{\sqrt{2}} & 0 & \frac{1}{2} & -\frac{1}{2} \\ 0 & \frac{1}{\sqrt{2}} & -\frac{1}{2} & -\frac{1}{2} \\ \frac{1}{\sqrt{2}} & 0 & -\frac{1}{2} & \frac{1}{2} \\ 0 & \frac{1}{\sqrt{2}} & \frac{1}{2} & \frac{1}{2} \end{bmatrix} \begin{bmatrix} 2 & 0 & 0 \\ 0 & \sqrt{2} & 0 \\ 0 & 0 & 0 \\ 0 & 0 & 0 \end{bmatrix} \begin{bmatrix} \frac{1}{\sqrt{2}} & 0 & \frac{1}{\sqrt{2}} \\ 0 & 1 & 0 \\ \frac{1}{\sqrt{2}} & 0 & -\frac{1}{\sqrt{2}} \end{bmatrix} = BDC.$$

This is an exact factorization, which could be obtained for any $A_{m \times n}$. We can consider an approximation by retaining some largest singular values and ignoring (making them 0) the smaller singular values. For example, here if we ignore $\sqrt{2}$, that is approximate $D$ to

$$D = \begin{bmatrix} 2 & 0 & 0 \\ 0 & 0 & 0 \\ 0 & 0 & 0 \\ 0 & 0 & 0 \end{bmatrix},$$

then the resulting approximation to $A$ based on the largest singular value is $A_1$ where

$$A_1 = \begin{bmatrix} 1 & 0 & 1 \\ 0 & 0 & 0 \\ 1 & 0 & 1 \\ 0 & 0 & 0 \end{bmatrix} = \begin{bmatrix} \frac{1}{\sqrt{2}} & 0 & \frac{1}{2} & -\frac{1}{2} \\ 0 & \frac{1}{\sqrt{2}} & -\frac{1}{2} & -\frac{1}{2} \\ \frac{1}{\sqrt{2}} & 0 & -\frac{1}{2} & \frac{1}{2} \\ 0 & \frac{1}{\sqrt{2}} & \frac{1}{2} & \frac{1}{2} \end{bmatrix} \begin{bmatrix} 2 & 0 & 0 \\ 0 & 0 & 0 \\ 0 & 0 & 0 \\ 0 & 0 & 0 \end{bmatrix} \begin{bmatrix} \frac{1}{\sqrt{2}} & 0 & \frac{1}{\sqrt{2}} \\ 0 & 1 & 0 \\ \frac{1}{\sqrt{2}} & 0 & -\frac{1}{\sqrt{2}} \end{bmatrix}.$$

Note that the squared Frobenius norm $||A - A_1||_F$ is 2 or the Frobenius norm is $\sqrt{2}$ which is the singular value that is ignored. It is not a coincidence. In general, if a matrix $A$ is approximated to $A_K$ by using the top $K$ singular values in $D$, then $||A - A_K||_F = \sigma_{K+1}^2$ where $\sigma_{K+1}$ is the largest of the ignored singular values. This helps in monitoring the possible error in approximating $A$ to $A_K$ for both dimensionality reduction and clustering. A popular application is in document representation, clustering, and classification under *latent semantic analysis (LSA)*.

**Table 4.1** Optimization in clustering and dimensionality reduction

| Specific task | Criterion function | Solution | Regularizer (Domain knowledge) |
|---|---|---|---|
| PCA | Minimize $E[(x^l - x^K)^t(x^l - x^K)]$ | Eigenvectors Covar. matrix | Best $K$ (Domain) |
| KMA | Minimize squared error | Local minimum | Diverse Centers |
| Hierarchical clustering | Minimum spanning Tree | Dendrogram of clusters | Dendrogram cut appropriately |
| MI based Feat. Selection $MI$ | Maximize features | $K$ Best | Consider all classes |
| SVD | $A = BDC$ | Exact | Approx. $A_K$ |

Under the matrix factorization, one can characterize any linear feature extraction including feature selection, hard and soft clustering, and even classification. Note that even nonlinear problems may be viewed as linear in an appropriate high-dimensional space. So, linear algebra in general and matrix factorization in particular are important in several of these topics.

Even the probabilistic variants like *probabilistic latent semantic analysis (PLSA)* are shown to be equivalent to deterministic factorization approaches like *NMF* and the *KMA*. This happens because both the approaches depend on some empirical schemes, based on the given data set in practice. In a more general sense *statistics* is responsible for the equivalence. An important semantic underlying matrix factorization is some kind of criterion function that is optimized with additional constraints to *regularize or reduce the diversity* of the solution space. We summarize the optimization related properties associated with clustering and dimensionality reduction in Table 4.1.

### 4.2.5 Diversified Clustering

Conventionally in clustering, the points in each cluster are similar to each other and points in different clusters are dissimilar. However, there are applications where each cluster needs to have diverse elements and a pair of clusters are highly similar. In other words there is a higher within cluster entropy and lower between cluster entropy.

Some of these applications are in

- *Peer Learning*: If a collection of students, selected based on some qualifying score, are to be grouped then the conventional clustering will lead to *stratified grouping*. In such a grouping all the students similar in terms of the qualifying score will be put together. This reduces the chance for peer learning. It can be shown to be good if each group has diverse students, that is students with varying qualifying scores. Further, to avoid discrimination between groups different groups should have similar collective behavior. This means round-robin allotment students to groups is a better deal than stratified grouping.

- *Team formation*: When different soccer teams are to be selected to participate in a cup, there will be diversity in terms of special roles of players like the goalkeeper, wing, center forward, full back, etc., This means there will be diversity in terms of these special roles in each team. Further, every team requires a goalkeeper, two wings, etc., which means a pair of teams are structurally similar. Not only in sports, this kind of grouping is required in the formation of committees and many other team formation scenarios.
- *Groups based on a Standard*: UG programmes offered by various computer science departments typically follow ACM curriculum. So, the similarity between different UG programmes exists because of the standard like the ACM curriculum. At the same time, each programme needs to show enough diversity in terms of representation of theoretical CS, computer systems, and other topics like ML, AI, DBMS, graphics, etc. There other standards like, for example, the Dewey Decimal Classification, Library Congress classification, etc. which are followed by libraries across the globe.

## 4.3  Classification

We have seen in earlier chapters how search and representation impact the classifiers. Knowledge is used in the form of prior densities, selection of representation schemes for patterns and classes. We can search for how knowledge can be exploited in modeling, selecting the correct model, and even selection of the values of the hyperparameters. Search takes different forms including searching for a solution to an optimization problem based on some constraints. In this section, we will examine how optimization can be used in modeling and selecting classifier models.

A good number of classifiers are explicitly modeled or can be interpreted as solutions to some *intuitively appealing and convenient* optimization problems. We will look into some of the classifiers.

### 4.3.1  Perceptron

It may be viewed as minimizing the sum of the violations of the training patterns, their distances from the wrong side of the decision boundary, using the current $w$, the weight vector of perceptron. This happens because $w$ has misclassified some training patterns. Noting that each such pattern, $x$ satisfies $w^t x < 0$, the perceptron criterion function based on $w$ is, $PCF(w)$ is

$$PCF(w) = - \sum_{x:w^t x < 0} w^t x.$$

$w^t x$ captures the extent of violation of $x$ because of $w$. Because $w^t x < 0$ for such an $x$, we minimize $-w^t x$ for every $x$ that is misclassified by $w$ so that sum of the extent of violations is minimized.

If we consider the gradient $\nabla_w PCF(w)$, we get $-\sum_{x:w^t x < 0} x$. So, if we use the gradient descent method to minimize $PCF(w)$, then the updates to $w$ are given, using the negative of the gradient with a suitable scaling factor $\eta$, by

$$w_{k+1} = w_k + \eta \sum_{x:w_k^t x < 0} x \qquad (4.1)$$

This update rule is called *batch mode update*. There are several simplifications to this equation.

1. One variant is to use $\eta = 1$ and consider one $x$ that is misclassified at a time rather than the sum of all the patterns $x$ that are misclassified by $w_k$. This is popularly called the *fixed increment rule* that we discussed in the previous chapter.
2. Another variant is to insist that the $w$ obtained is a simple sparse vector, minimum possible nonzero entries, that can be effectively used for classification which is useful in high-dimensional spaces. This is specified by

$$PCF(w) = - \sum_{x:w^t x < 0} w^t x + \lambda' w^t w, \qquad (4.2)$$

so that while minimizing the sum of violations, we reduce the nonzero entries in $w$ as well. There is a scaling factor $\lambda'$. Noting that the gradient of $w^t w$ is $2w$, we have the the the corresponding *incremental update rule*, one pattern at a time, to be

$$w_{k+1} = w_k + \eta x^k - \lambda w_k \rightarrow w_{k+1} = (1 - \lambda) w_k + \eta x^k$$

where $\lambda = 2\eta\lambda'$ and $x^k$ is the first pattern misclassified by $w_k$.

Note that both these variants are constraining or regularizing the optimization solution, $w$.

## 4.3.2 Support Vector Machine (SVM)

In SVMs, the criterion function that is considered is *margin between the two classes*. This may be detailed using Fig. 4.1. In *SVM* margin between the positive and negative classes is maximized. In the figure, there are negative class patterns in the left side. These are labeled by using $-$. Similarly, on the right side we have the positive class patterns. These are labeled by $+$.

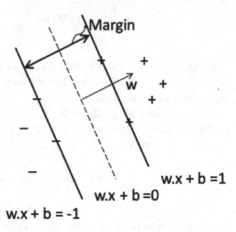

**Fig. 4.1** Margin between
the two classes

In this two-dimensional case, there are two parallel lines (in higher dimensions they will be parallel hyperplanes) called *support lines*. The respective class *boundary patterns* are located on these support lines. The negative class patterns satisfy the property that $w^t x + b \leq 1$ where the boundary vectors, $x$s, or *support vectors (SVs)* of the negative class satisfy $w^t x + b = -1$. Similarly, the positive class patterns satisfy $w^t x + b \geq 1$ with the respective $SVs$ satisfying the property $w^t x + b = +1$.

The decision boundary between the two classes is characterized by points $x$ such that $w^t x + b = 0$. Points to its right are from positive class and left side patterns are of negative class. If two points $x_1$ and $x_2$ are points on the decision boundary, then

$$w^t x_1 + b = w^t x_2 + b = 0 \Rightarrow w^t(x_1 - x_2) = 0.$$

This means vector $w$ is orthogonal to $x_1 - x_2$ or the line on which they are located which is the decision boundary itself. So, $w$ is orthogonal to the decision boundary as shown in the figure.

Another property is that $w$ points towards the positive side. Consider a problem where the origin is on the decision boundary. So, $w^t 0 + b = 0 \Rightarrow b = 0$. Now if we consider a point $x_1 \in c_+$ the positive class, then $w^t x_1 > 0$. The cosine of the angle, $\theta$, between $w$ and $x_1$ is given by

$$cos\theta = \frac{w^t x_1}{||w|| ||x_1||}.$$

The denominator terms are positive here and the numerator is positive as $x_1 \in c_+$. So, $cos\theta > 0 \Rightarrow$ the angle between $w$ and $x_1$ is acute. So, $w$ points towards the positive side.

Any point $x \in c_+$ may be written as $x = x_d + p\frac{w}{||w||}$ where $x_d$ is point on the decision boundary at which the normal projection of $x$ onto the decision boundary meets it. If the distance between $x$ and $x_d$ is $p$ units, then the corresponding vector is

$p\frac{w}{||w||}$ because $w$ is orthogonal or normal to the decision boundary. But as $x \in c_+$,

$$w^t x + b = w^t(x_d + p\frac{w}{||w||}) + b = w^t x_d + b + p||w|| = p||w|| > 0$$

as $w^t x_d + b = 0$, where $\frac{w}{||w||}$ is a unit vector in the direction of $w$. So,

$$w^t x + b = p||w|| \Rightarrow p = \frac{w^t x + b}{||w||}.$$

So, normal distance between any point $x$ on the positive support line and the decision boundary is $\frac{w^t x + b}{||w||} = \frac{1}{||w||}$. Similarly, from any point $x$ on the negative support line to the decision boundary the modulus of the distance is again $\frac{1}{||w||}$. So,

$$margin = \frac{1}{||w||} + \frac{1}{||w||} = \frac{2}{||w||}.$$

In $SVM$, we find $w$ that *maximizes the margin*. Equivalently, we minimize $\frac{1}{2}||w||^2$ which maximizes the margin. The constraints are $y_i(w^t x_i + b) \geq 1$ where $y_i$ is the class label of $x_i$; $y_i = 1 \ or \ -1$ based on whether $x_i \in c_+$ or $x_i \in c_-$ respectively.

We can express the corresponding Lagrangian by taking into account the constraints as

$$\mathcal{L}(w, b, \alpha) = \frac{1}{2}||w||^2 + \sum_{i=1}^{n} \alpha_i(1 - y_i(w^t x_i + b)),$$

where we would like to find the vectors $w, \alpha = \{\alpha_1, \ldots, \alpha_n\}$ and the scalar $b$. Optimal values of these variables can be obtained by equating the gradient to zero which is given by

$$\nabla_w \mathcal{L} = w - \sum_{i=1}^{n} \alpha_i y_i x_i = 0 \Rightarrow w = \sum_{i=1}^{n} \alpha_i y_i x_i.$$

$$\nabla_b \mathcal{L} = \sum_{i=1}^{} \alpha_i y_i = 0.$$

$$\nabla_{\alpha_i} \mathcal{L} = 1 - y_i(w^t x_i + b) = 0 \Rightarrow y_i(w^t x_i + b) = 1 \Rightarrow w^t x_i + b = y_i$$

There are other conditions also including $\alpha_i \geq 0$ and $\alpha_i(1 - y_i(w^t x_i + b)) = 0$.

By using these equations all the variables $w$, $\alpha$, and $b$ can be determined using different approaches. Here, the optimization problem was chosen such that it is a well-behaved problem guaranteeing a globally optimal solution to the minimization. However, we face a difficulty when there is no margin between the two classes. This can happen, for example in Fig. 4.1, if a point from the positive class falls to the left of the decision boundary or equivalently a point from the negative class falls to the right of the decision boundary. Such points are called *violators*. This can be the case in most of the real-world problems.

To overcome this problem, a popularly used solution is to formulate it as a *soft margin problem*. This is achieved by weighing each of the violators using a weight $C$ based on the extent of violation. If we do not want to permit any violator then $C \to \infty$. This amounts to the soft margin formulation to converge to the hard margin formulation. On the other extreme, a value of $C = 0$ means every point can be a violator. However, this will not solve the problem in practice.

Typically, a positive finite nonzero value is used for $C$ to accommodate some violators. The corresponding problem is

$$\min_{w} \frac{1}{2}||w||^2 + C \sum_{i=1}^{n} \xi_i .$$

*s.t.* $y_i(w^t x_i + b) \geq 1 - \xi_i$. and $\xi_i \geq 0$. It is seen that there is no change in the form of the variables. Only change is that in the hard margin formulation, $\alpha_i \geq 0$. In the soft margin case, $0 \leq \alpha_i \leq C$.

An important practical consideration is the right value of $C$. This shifts the attention from getting the global optimal solution to getting the right value of $C$. So, tuning of hyperparameter $C$ occupies the central stage in practice. Some professionally developed software packages have helped in realizing this practically.

### 4.3.3  Summary

In this chapter, we have seen the role of optimization in dimensionality reduction, clustering, and classification. We have considered only some of the algorithms. There are potentially a large variety of other machine learning platforms like neural networks. In a sense optimization based solutions exhibit diversity which is controlled using regularization to provide more central or less variance solutions.

Note the following about optimization. The set of constraints specify the *feasible region*. This may typically characterize potentially infinite solutions or diverse possible solutions. The criterion function being optimized will force the selection of one or more of these diverse points in the solution space increasing the centrality. A regularizer will shrink this collection of possible solutions further.

Consider, for example, a data set of the following four patterns drawn from 2 classes as shown in Table 4.2. Let patterns 1 and 2 be training points from class 1 and let the other 2, that is patterns 3 and 4 be from class 2.

Let a classifier gave two $w$ vectors given by
$w^1 = (1, 0, 1, 0, -2)$
$w^2 = (0.5, 0.6, 0.5, 0.4, -2)$. Verify that both these weight vectors classify all the four patterns correctly. For example, using $w^1$ on pattern 2 gives us

**Table 4.2** 4-dimensional data from two classes

| Pattern | $x_1$ | $x_2$ | $x_3$ | $x_4$ |
|---------|-------|-------|-------|-------|
| 1 | 0.6 | 0.4 | 0.7 | 0.4 |
| 2 | 0.5 | 0.3 | 0.7 | 0.5 |
| 3 | 1.2 | 1.4 | 1.5 | 1.6 |
| 4 | 1.3 | 1.3 | 1.4 | 1.5 |

$0.5 + 0 + 0.7 + 0 - 2 < 0$. Similarly $w^2$ with pattern 4 gives us $0.65 + 0.78 + 0.7 + 0.9 - 2 > 0$. Similarly one can verify other patterns.

Among these two weight vectors, if we require a sparse vector, then $w^1$ will be selected and $w^2$ will be left out.

# Bibliography

1. Manning CD, Raghavan P, Schütze H (2008) Introduction to information retrieval. Cambridge University Press
2. Fan R-E, Chang K-W, Hsieh C-J, Wang X-R, Lin C-J (2008) LIBLINEAR: a library for large linear classification. JMLR 9:1871–1874
3. Hsu CW, Lin C-J (2002) A comparison of methods for multiclass support vector machines. IEEE Trans Neural Netw 13(2):415–425
4. Lay DC (2012) Linear algebra and its applications. Addison-Wesley
5. Murty MN, Devi VS (2015) Introduction to Pattern recognition and machine learning. IISc Press
6. Ding C, Li T, Peng W (2008) On the equivalence between Non-negative Matrix Factorization and Probabilistic Latent Semantic Indexing. Comput Stat Data Anal 52:3913–3917
7. Chaudhuri AR, Murty MN (2012) On the relation between K-means and PLSA. In: Proceedings of ICPR, Nov 11–15 2012: pp 2298–2301, Japan
8. Duda RO, Hart PE, Stork DG (2001) Pattern classification. Wiley-Interscience
9. Sambaran B, Sharad N, Rishabh D, Murty MN (2018) DivGroup: a diversified approach to divide collection of patterns into uniform groups. ICPR 2018:964–969
10. Rakesh A, Sharad N, Murty MN (2017) Grouping students for maximizing learning from peers. EDM

# Chapter 5
# Ranking

**Abstract** Ranking is an important task in machine learning, information retrieval, and data mining. We consider different notions like similarity and density and their role in ranking. Further, we discuss how centrality and diversity are captured in a variety of ranking tasks.

**Keywords** Similarity · Search engine · Centrality · Diversity

## 5.1  Introduction

We have seen roles of centrality and diversity in several AI tasks including representation, clustering, and classification. Another important task is *ranking*. Ranking deals with assigning each point in a collection, $\mathcal{X} = \{x_1, \ldots, x_K\}$ to an integral value in the range $[1, 2, \ldots, K]$. In its simplest form, it may be viewed as classification where we bin items into two groups that are ranked. There are several applications of ranking including presenting results of a search engine against a user query.

## 5.2  Ranking Based on Similarity

This is prominently used in the context of search engines. Here, the problem is given a collection of documents, $\mathcal{D} = \{d_1, \ldots, d_n\}$ and a user query, $q$, to rank the matching documents based on similarity. Here, query also is viewed as a document. In a simple content-based retrieval, we represent each document as a vector of size $|V|$ where $V$ is the set of vocabulary. If $|V| = l$, that is the number of words in the vocabulary is $l$, then $q$ and any document $d_i$ are represented as $l$-dimensional vectors. The entry in each of the $l$ locations is the corresponding $TF - IDF$ value. Now, similarity between $q$ and $d_i, i = 1, \ldots, n$ is typically computed by using the cosine of the angle between $q$ and $d_i$ which is given by

M. N. Murty and A. Biswas, *Centrality and Diversity in Search*,
SpringerBriefs in Intelligent Systems, https://doi.org/10.1007/978-3-030-24713-3_5

$$cos(angle(q, d_i)) = \frac{q^t d_i}{||q|| ||d_i||},$$

where the denominator terms in the $RHS$ are the Euclidean norms of $q$ and $d_i$.

Note that smaller the angle between $q$ and some $d_j$ larger the cosine value or equivalently higher the similarity between $q$ and $d_j$. This observation helps us to rank documents against a $q$ by ordering the documents in the decreasing order of similarity with query $q$.

It was observed that similarity computation based on content alone is inadequate to make an effective retrieval of web documents. It was proposed to include the (hyper-)link structure that is present among the web pages. Specifically, importance of a web document is a weighted combination of the content match with the query and centrality of the web page with which it is associated. The centrality of a web page, in a simple manner, is a recursive characterization based on a weighted combination of centrality of pages linked or connected to it. For example, the eigenvector associated with the largest eigenvalue of the adjacency matrix of the web graph provides a recursive characterization of centrality of all the nodes in the network.

## 5.3   Ranking Based on Density

In applications like clustering based on centroids/representatives, it is important to select an appropriate set of initial cluster representatives. We have seen that such representatives need to be diverse. However, outlier points are from low-density regions. So, in the process of maintaining diversity, outlier points should not be considered as centroids. A balance between centrality and diversity can be maintained by selecting representative patterns from *dense regions*. So, we need to rank the possible centroids based on density and ensure diversity by selecting an appropriate subset of this ranked set of dense points.

When a search engine displays results against a query, one possibility is to identify density with a concept that is popular among a good number of documents. So, each concept/cluster of documents is associated with the cardinality of the cluster, which can be viewed as the density. We rank these concepts by ranking based on density in terms of size of the associated clusters.

Ranking based on density is useful in several other applications like facility location, transportation arrangements, among others. Here, density is associated with the number of possible customers depending on a facility from a location. Density need not be simply based on location alone; it could also be based on time, and other factors.

## 5.4  Centrality and Diversity in Ranking

Searching a collection of documents against a user query and displaying the results in a ranked order for the consumption of the user is an important activity in information retrieval. Such a task is routinely performed by *search engines*.

However, in performing this task search engines consider documents as influential/central if they match the query and are associated with influential websites. The statistics about users scrolling the search engine results indicate that a large percentage of users never scroll beyond the first page that typically contains less than a dozen results. So, it is likely that out of a large number of results displayed against a query, a vast majority of the results go unnoticed by the users.

So, it is practically important that the first page of results has enough diversity against the query so that an average user is satisfied. For example, consider the query word *model*, which will bias the search engine toward the popular usage of the word in the fashion industry. Documents related to *model of the automobile* or models as in hidden Markov models may fail to show up in the first page of results displayed by the search engine. So, the challenge is to present results on the first page by diversifying them to reflect all these different meanings of the same term.

### 5.4.1  Diversification Based on a Taxonomy

If a knowledge structure/taxonomy capturing the different relevant topics is available, then it is possible to exploit the taxonomy, that is a hierarchical structure describing various categories, to provide diverse search results. If relevance/centrality alone is considered to rank and display the results then diversity could get sacrificed. So, a judicious mix of results based on relevance and diversity can be achieved if taxonomy also is exploited where both the documents and the query may be associated with more than one subtree in the hierarchy.

Not just in terms of ranking and displaying the results, even the evaluation of search engines must take into consideration the diversity among the results instead of relevance alone.

## 5.5  Ranking Sentences for Extractive Summarization

In summarizing documents, there are two categories:

1. *Abstractive summarization*: Here, some natural language understanding capability is required of the system to abstract the important semantic concepts in the document. A summary of the document is generated by well-formed sentences based on these concepts and interactions between them.

2. In *extractive summarization* of a text document, a learning algorithm is typically employed to rank sentences based on some criterion. For example, the presence of high-frequency words, and location of sentences can be used to rank the sentences. A combination of these important/central sentences is used to form the summary.

   Another possibility is to rank a sentence in the document based on entropy of the sentence. Here, Shannon's entropy is computed based on estimating the probabilities using the collection frequencies of words in the document; the collection frequency of a word is the frequency of the word in the document collection. It is possible that ranking words based on entropy alone may not be adequate. If the training data consists of labeled documents, then one can use topic models like *Latent Dirichlet Allocation (LDA)*. This will help in identifying summary topics in a document and exploit them in extracting sentences still maintaining the required diversity.

## 5.6   Diversity in Recommendations

Typically while making recommendations to a user, a problem that is encountered is that important/central recommendations are made. Because of the lack of diversity in the recommendations, an average user may not be satisfied as the set of recommendations is not comprehensive.

Conventionally, diversity is captured with the help of distances between items that are represented as vectors based on their attributes. For example, in K-means++, diverse initial centroids are obtained based on interpoint distances. There are some limitations to this approach.

1. It is possible that there is no semantic information available with every item considered for possible recommendation.
2. Even if the semantic information is available, it may be difficult to capture diversity through the semantics. For example, two different books authored by the same person may not be similar.

However, it is possible to bring in a regularizer based on entropy to diversify recommendations. Instead of pairwise distance/similarity, it can capture the gestalt information provided by the entire vector space. It is shown that when the items recommended are orthogonal in their vector space representation, then the entropy regularizer will be maximized indicating the presence of diversity between the item vectors. On the contrary, if the item vectors are linearly dependent, then the entropy regularizer is minimized indicating the absence of diversity between the items.

Another way to ensure diversity in recommendations is based on employing a non-Markovian random walk where the transition probability matrix is not static, rather it can change with time. One can employ a Vertex-Reinforced Random Walk (VRRW). It is based on the idea that the future transition probability, from node $v_i$ to node $v_j$, is influenced by the number of times node $v_j$ has been visited in the past.

This will be justified because even though two competing books or movies start with a similar size customer base, over a period of time one of the items (book/movie) may become more popular and hence may increase its customer base. This can impact the diversity in the resulting recommendations because if a node $v_j$ has a higher transition probability, from $v_i$, for the random walk to visit it, similar nodes that are immediately connected to $v_j$ will attract lesser number of visits by the random walk under the time-varying transition probabilities.

## 5.7 Summary

In this chapter, a detailed discussion on ranking was provided. Specific topics included were on ranking based on similarity and density. The roles of centrality and diversity in ranking were considered. Specifically, diversification of ranking results based on taxonomies, in recommendations, and in extractive summarization of documents was considered.

## Bibliography

1. Manning CD, Raghavan P, Schütze H (2008) Introduction to information retrieval. Cambridge University Press
2. Witten IH, Frank E, Hall MA (2011) Data mining, 3th edn, Morgan Kauffmann
3. Agrawal R, Sreenivas G (2009) Diversifying search results, WSDM 2009, Barcelona
4. Karthik N, Murty MN (2012) Obtaining single document summaries using latent Dirichlet allocation, ICONIP 2012. LNCS 7666
5. Qin L, Zhu X (2013) Promoting diversity in recommendation by entropy Regularizer, IJCAI 2013, Beijing
6. Sharad N, Aayush M, Murty MN (2018) Fusing diversity in recommendations in heterogeneous information networks, WSDM 2018, Marina Del Rey

# Chapter 6
# Centrality and Diversity in Social and Information Networks

**Abstract** There are several applications where centrality and diversity can play important roles. These include networks and recommendation systems. We deal with networks in this chapter. We specifically examine the representation of networks using graphs, centrality in social networks, and diversity among communities in benchmark network data sets.

**Keywords** Social networks · Structure · Content · Centrality · Diversity

## 6.1 Introduction

Networks are gaining their importance and popularity in AI, machine learning, and a variety of other tasks. In any of our day-to-day applications, it is important to view the collection of underlying entities, not as a set of isolated objects/events, but as a single network of entities where different subsets of the collection satisfy one or more relationships. This abstraction permits us to exploit the domain knowledge in the best possible way to analyze and understand the underlying dependencies among the entities. Graph is the most popularly used structure to represent networks. We discuss the representation of networks using graphs after introducing the related notions.

A *network* is a structure made up of a set of possible links between collections of nodes. For different applications, these notions *node* and *link* could stand for different entities.

For example, in *citation networks*, each node represents a publication and a link from node $n_i$ to node $n_j$ indicates the publication corresponding to $n_i$ has cited the publication behind $n_j$. In *road networks* each city is represented by a node and a link between $n_i$ and $n_j$ represents existence of a road between the two cities.

**Fig. 6.1** Example network

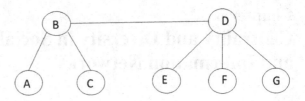

## 6.2  Representation

A network is popularly represented as a simple graph. We illustrate it with the help of an example shown in Fig. 6.1.

Some important observations from the figure are as follows:

1. There are *seven nodes* in the example network; they are labeled $A$ to $G$.
2. There is a link/edge between some pairs of nodes; for example, node pairs $A$, $B$ and $G$, $D$.
3. Edges are absent between some pairs of nodes; for example, node pairs $A$, $F$ and $F$, $G$.
4. The graph shown in Fig. 6.1 is *undirected*. The link between $D$ and $G$ may be represented either by $(D, G)$ or $(G, D)$; there is no difference. Such a representation abstracts *association* between the two nodes. For example, *A has coauthored a publication with B* is the same as *B has coauthored a publication with A*. Such relations that are symmetric can be captured by edges/links that are *undirected* in the graph.
5. A *path* between a pair of nodes $A$ and $B$ is a sequence of distinct nodes $A, n_1, n_2, \ldots, n_{l-1}, B$ such that there exists an edge between two successive nodes in the sequence. The length of the path or *Path Length* is $l$ if there are $l + 1$ nodes in the path.
   There is a path of length 3 between $A$ and $E$; the corresponding sequence is $A$, $B$, $D$, $E$. Similarly, there is a path of length 2 between $B$ and $F$ specified by $B$, $D$, $F$.
6. The number of edges associated with a node is the *degree* of the node in an *undirected graph*. The degree of node $B$ is 3 and the degree of $D$ is 4.
7. In a *directed graph*, the degree is split into *indegree* and *outdegree*.

   - The *indegree* of a node $A$ is the number of edges pointing to $A$, that is edges of the form $(x, A)$, for some node $x$. In a directed graph, $(x, A)$ is not the same as $(A, x)$.
   - Similarly, *outdegree* of $A$ is the number of edges going out of $A$, that is edges of the from $(A, x)$.

## 6.3 Matrix Representation of Networks

There are different types of networks that require different representation schemes. Some of them may require schemes based on *hypergraphs*. However, popularly known networks are *represented using graphs*. Further, graphs are typically represented on the machine using either the *adjacency matrix* representation or the *adjacency list* representation. We consider the adjacency matrix here because of its popularity.

- *Adjacency Matrix (A)*: The adjacency matrix of a graph is a *square matrix A* of size $n \times n$ where $n$ is the number of nodes in the graph. The $(i, j)$th entry, $A_{ij}$, in $A$ is 1 if there is an edge between nodes $i$ and $j$; if there is no link then the entry is 0. If the *graph is undirected*, then the adjacency matrix is *symmetric* also. The adjacency matrix corresponding to the undirected graph in Fig. 6.1 is given in Table 6.1.
- **Number of Paths of Length 2:**
  The adjacency matrix $A$ of a graph abstracts edges or paths of length 1. We get paths of length 2 by considering the matrix $A^2(A \times A)$. The matrix $A^2$ for the example graph in Fig. 6.1 is given in Table 6.2
- Note that in $A^2$ the $i$th diagonal entry, $A_{i,i}$ corresponds to the degree of the $i$th node.
- Note that some entries in $A^2$ are 0 indicating that there are no paths of length 2 between the corresponding pair of nodes. For example, the entry for node pair $B, C$ is 0 meaning that there are no paths of length 2 between $B$ and $C$.
- Observe that every non-diagonal entry is either 0 or 1. This is the case because of the simplicity of the example in Fig. 6.1.
- The number of paths of length 2 between $B$ and $F$ is 1. So, there is a third node on the path between $B$ and $F$ which happens to be $D$ in this example. Such a node is a neighbor of both $B$ and $F$ and is called a *common neighbor*.

Some of the basic notions *path*, *common neighbors*, and *degree* is useful in analysis and prediction tasks associated with networks which we will consider in detail later.

## 6.4 Link Prediction

*Link prediction (LP)* is one of the important prediction tasks in networks. A network is typically viewed as a set $G = \{V, E\}$ where $V$ is the set of vertices of some type and $E$ is the collection of edges.

The LP problem is specified as follows: Given a static snapshot of the network at time $t$ (that is $G_t = \{V_t, E_t\}$ is given) to predict the edge set at time $t + 1$, $E_{t+1}$, assuming that $V_{t+1} = V_t$.

**Table 6.1** Adjacency matrix for the graph in Fig. 6.1

| Node/Node | A | B | C | D | E | F | G |
|---|---|---|---|---|---|---|---|
| A | 0 | 1 | 0 | 0 | 0 | 0 | 0 |
| B | 1 | 0 | 1 | 1 | 0 | 0 | 0 |
| C | 0 | 1 | 0 | 0 | 0 | 0 | 0 |
| D | 0 | 1 | 0 | 0 | 1 | 1 | 1 |
| E | 0 | 0 | 0 | 1 | 0 | 0 | 0 |
| F | 0 | 0 | 0 | 1 | 0 | 0 | 0 |
| G | 0 | 0 | 0 | 1 | 0 | 0 | 0 |

**Table 6.2** Square of the adjacency matrix for the graph in Fig. 6.1

| Node/Node | A | B | C | D | E | F | G |
|---|---|---|---|---|---|---|---|
| A | 1 | 0 | 1 | 1 | 0 | 0 | 0 |
| B | 0 | 3 | 0 | 0 | 1 | 1 | 1 |
| C | 1 | 0 | 1 | 1 | 0 | 0 | 0 |
| D | 1 | 0 | 1 | 4 | 0 | 0 | 0 |
| E | 0 | 1 | 0 | 0 | 1 | 1 | 1 |
| F | 0 | 1 | 0 | 0 | 1 | 1 | 1 |
| G | 0 | 1 | 0 | 0 | 1 | 1 | 1 |

Note that $E_{t+1} - E_t$ is the set of missing links that need to be predicted. Typically an LP algorithm computes the similarity between all pairs of nodes that are not yet linked using the graph $G_t$ and ranks the edges based on the computed similarity values.

### 6.4.1 LP Algorithms

These LP algorithms are categorized into *local* and *global* LP algorithms based on whether the similarity is computed using the local neighborhood of the two end vertices or a global neighborhood is used. We provide some related details next.

- *Local Similarity functions*: Let $A$ and $B$ be the two end vertices of the link to be predicted and let $ND(A)$ and $ND(B)$ be the sets of neighboring nodes of $A$ and $B$ respectively. Then

  1. *Common-neighbors(A,B)* $= |ND(A) \cap ND(B)|$.

  The common neighbors ($CN$) based similarity function is the simplest example of local computation of similarity. Here, all the common neighbors are given *equal importance* of 1 unit each. Note that

$$CN(A, B) \in [0, min(|ND(A)|, |ND(B)|)].$$

2. $Jaccard\text{-}coefficient(A,B) = \frac{|ND(A) \cap ND(B)|}{|ND(A) \cup ND(B)|}$.

The CN function does not normalize the score. Consequently, the score can be large if $ND(A)$ and $ND(B)$ are large size sets and it may assume smaller values if $ND(A)$ and $ND(B)$ are smaller in size.

The Jaccard coefficient($JC$) similarity function *normalizes the CN score*, between a pair of nodes, by dividing it with the size of the union of the sets $ND(A)$ and $ND(B)$. Note that $JC(A, B) \in [0, 1]$.

3. $Adamic\text{-}Adar\ score(A,B) = \sum_{x \in ND(A) \cap ND(B)} \frac{1}{\log |ND(x)|}$.

Here, each $CN$ contributes differently based on the logarithm of the degree of the $CN$. The minimum degree of any $CN$ node is 2 because it is linked to both $A$ and $B$, perhaps in addition to other possible nodes. So, if the base of the logarithm is 2, then $\log_2 |ND(x)| \geq 1$.

The minimum possible Adamic–Adar($AA$) value occurs when the numerator has the minimum possible value of 1 and the denominator has the maximum possible value which is $\log_2(n - 1)$, where there are n nodes in the graph. In such a case, the value will be $\frac{1}{\log_2(n-1)}$. The maximum value occurs when the numerator has maximum value of $n - 2$ and the denominator has a minimum value of $1 = \log_2 2$. So, $AA(A, B) \in [\frac{1}{\log_2(n-1)}, (n - 2)]$.

4. $Preferential\text{-}attachment\text{-}score(A,B) = |ND(A)| \times |ND(B)|$

It promotes edges/links between high-degree nodes. For example, in preferential attachment($PA$), the minimum possible value of $PA(A, B)$ is 1 when $A$ and $B$ have degree 1 each. The maximum value of $PA$ is $(n - 1)^2$ which occurs when $A$ and $B$ has a maximum possible degree of $n - 1$. So, $PA(A, B) \in [1, (n - 1)^2]$.

We have indicated a small list of local $LP$ similarity functions. There are several others.

- *Global Similarity functions*: We consider two global similarity functions.

  - *Graph Distance*: Here, the similarity, $gds$, between a pair of nodes $A$ and $B$ is inversely proportional to the length of the shortest path between $A$ and $B$.

  $$gds(A, B) = \frac{1}{length\ of\ the\ shortest\ path(A, B)}.$$

  - *Katz's Similarity* is a popular global similarity function that is based on paths between the two nodes. If $pl_i$ is the number of paths of length $i = 2, \ldots, q$ between the nodes $A$ and $B$, then

$$katz - similarity(A, B) = \sum_{i=2}^{q} \beta^i pl_i,$$

where $\beta$ is a parameter that assumes typically a value in the range $(0, 0.1]$. So, paths are combined here to arrive at the similarity.

## 6.5   Social and Information Networks

A *social network* is a network where the nodes represent *humans* and the links/edges characterize interactions/socialization among humans. An example interaction is the *friendship* which is the property characterized by a pair of nodes in the network and this property is *symmetric* as *A is a friend of B if B is a friend of A*. It is possible to represent a social network as a graph. In an abstract sense, a graph may be characterized using a *set of nodes* ($V$), a *set of edges/links* ($E$), and a *set of weights* ($W$). So,

- $G = \{V, E, W\}$ where
- $V = set\ of\ nodes, \{v_1, v_2, \ldots, v_n\}$
- $E = set\ of\ edges\ e_{i,j} \in E$ is the edge between $v_i$ and $v_j$ and
- $W = set\ of\ weights$

In a simple representation, we can have weight $w_{i,j} = 1$ if there is an edge between nodes $v_i$ and $v_j$; if there is no edge, then $w_{i,j} = 0$. This corresponds to a *binary representation* that characterizes the *presence* or *absence* of an edge between pairs of nodes. It is possible to have a more general representation where $w_{i,j} \in \Re$; here $\Re$ is the set of real numbers. However, we deal with only binary representation in this chapter. In such a case, we can simplify the notation and view the graph $G$ as

- $G = < V, E >$ where
- $V = set\ of\ nodes, \{v_1, v_2, \ldots, v_n\}$
- $E = set\ of\ edges\ present$; here edge $e_{i,j} \in E$ if there is an edge between nodes $v_i$ and $v_j$, else $e_{i,j} \notin E$.

It is convenient to extend the notion to entities other than humans as the resulting networks share a good number of interesting and useful properties. These networks include

- *Citation networks* Each node represents a paper $P$ and an edge between a pair of nodes, $P_i$ and $P_j$ abstracts a citation which is a directed edge.
- *Coauthor Networks* Here each author is represented by a node and a link between two nodes if the corresponding authors have coauthored a paper. This is represented by an undirected graph as *coauthorship* is symmetric.
- *Homogeneous and Heterogeneous Networks* In a *homogeneous network*, all the nodes are of the same type. For example, *friendship network* is homogeneous. On the other hand, in a *heterogeneous network* nodes could be of different types. For

example, in an academic network, nodes represent both the authors and papers. Further, the links could be of different types; *coauthorship* between two authors, *author of* between an author and a paper, and *cited by* link between two papers.
- *Information Networks* In addition to the structural information, attribute information is available with each node in several applications including coauthor networks.

We consider the analysis of only homogeneous social networks in this chapter.

## 6.6 Important Properties of Social Networks

We briefly discuss some of the important properties of a social network.

- *Power-Law Degree Distribution*: It is observed that *degree distribution* follows a *power law* asymptotically. Specifically, $N(i)$, number of nodes of degree $i$ is given by

$$N(i) \propto i^{-\alpha},$$

where $\alpha \in [2, 3]$ based on empirical studies. It is called *scale-free* because the form of $N(i)$ does not change with different scales for $i$. A plot of the $log\,N(i)$ versus $log\,i$ and $N(i)$ versus $i$ is shown in Fig. 6.4. Note that even if we scale $i$ by some positive real number $c$, the values on the $X$ axis will shift by a constant, that is by $log(c)$; this is because $log(ci) = log(c) + log(i)$. Hence the specific form of the plot will not change but for a parallel shift characterized by $log(c)$.
It means that in a given network, there will be a large number of *low degree nodes* and a small number of *high-degree nodes*.
This property is exploited in the *analysis of social networks*. For example, the *AA* similarity function gives less importance to high-degree *CN*s and larger importance to middle and low degree *CN*s. Note that this is similar to the power-law distribution of the frequencies of the terms in a collection of documents where high-frequency words are not discriminative.
- *The Small-World Phenomenon*: It is based on the observation that average path length between a pair of nodes is small. This is also called as *six degrees of separation* where it was observed that the median path length between a pair of nodes is 6 (Fig. 6.2).
This property is useful in the analysis that involves lengths of the paths between a pair of nodes. For example, in computing the *Katz* similarity, the value of $q$, that is the maximum path length is capped at a value of 6.
- *Exhibits Community Structure*: Intuitively we may say that nodes in $V^c \subseteq V$ are all in the same *community* if they are all similar to each other; or equivalently every pair of nodes in $V^c$ are similar. This is formally characterized using the notion of *clustering coefficient($CC$)*, which is defined as follows:

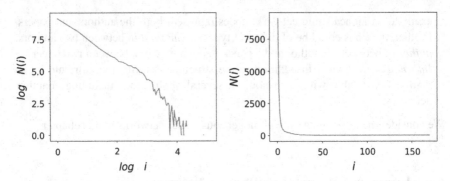

**Fig. 6.2** Power-Law Degree Distribution exhibited by Pubmed data set available at https://linqs. soe.ucsc.edu/data. This data set has 19717 nodes and 44327 edges. The maximum, minimum, and average degree of the data set is 171, 1, and 4.50 respectively

$$CC(v_i) = \frac{2\,|\,E_{v_i}\,|}{degree_i\,(degree_i - 1)}$$

Here, $degree_i$ is the degree of $v_i$. The maximum possible number of edges in the neighborhood of $v_i$ is $\frac{degree_i(degree_i-1)}{2}$ and $E_{v_i}$ is the number of edges present among the neighbors of $v_i$. Note that $CC_i$ has the maximum value of 1. This happens when every pair of neighbors of $v_i$ are connected by an edge.

The clustering coefficient of a cluster (or a subgraph) is the average of the clustering coefficients of all the nodes in the cluster (or the subgraph).

It is observed that for different types of networks, the clustering coefficient is reasonably larger than what random chance permits. Specifically, it is prominent in *Actor Networks*, *Metabolic Networks*, and *Coauthor Networks*.

## 6.7 Centrality in Social Networks

Centrality is a well studied and reasonably well-understood notions in networks. A node is *central or influential* if it is active or important in some sense. Some popular centrality characterization is based on degree, closeness, betweenness, and eigenvector.

### 6.7.1 Degree Centrality

This is the simplest notion of centrality. A node is more central if it can be reached by other nodes better. So, higher the degree of a node, larger is its centrality. We

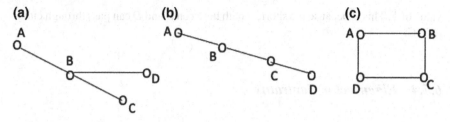

**Fig. 6.3** Three different graphs with four nodes

illustrate it with the help of Fig. 6.3. For the sake of simplicity, the four nodes are labeled $A, B, C$ *and* $D$ in all the three subfigures labeled (a), (b), and (c).

In figure (a), degree(B) is 3 and the remaining three nodes have degree 1 each. In figure (b), degree(B) = degree(C) =2 and the remaining two nodes have degree 1 each. In figure (c), each of the four nodes is of degree 2. So, node B in figure (a) is more central than any other nodes in figure (a) and any node in figures (b) or (c).

### 6.7.2 Closeness Centrality

It is defined for each node in a network.

$$Closeness - centrality(v_i) = \frac{1}{\sum_{j=1, j \neq i}^{n} d_{ij}}$$

where $d_{ij} = Shortest - path - length\ between\ v_i\ and\ v_j$ This means if a node $v_i$ closer to rest of the nodes in a network, then it has a high closeness centrality.

In Fig. 6.3a, node $B$ has a closeness centrality value of $\frac{1}{3}$. The other three nodes have a value of $\frac{1}{5}$ each. In figure (b), nodes $B$ and $C$ have a closeness centrality value of $\frac{1}{4}$ each and the remaining two nodes have $\frac{1}{6}$ each. In figure (c), every node has a closeness centrality value of $\frac{1}{4}$.

### 6.7.3 Betweenness Centrality

Betweenness centrality of a node $v_i$ is defined as the number of shortest paths between pairs of other vertices on which $v_i$ is located.

In Fig. 6.3a, node $B$ is on the shortest paths between $A$ and $C$; $A$ and $D$; and $C$ and $D$. So, its betweenness centrality value is 3. The other three nodes have a value of 0 each. In figure (b), nodes $B$ and $C$ have a betweenness centrality value of 2 each and the other two nodes have 0 each. In figure (c), each node has a betweenness centrality value of $\frac{1}{2}$. This is because the shortest path between $B$ and $D$ can pass through either $A$ or $C$.

### 6.7.4   Eigenvector Centrality

Eigenvector centrality is a recursive characterization of the centrality of nodes. The recursive specification is

$$e_i = \frac{1}{\lambda} \sum_{j=1}^{n} ad_{ij} e_j,$$

where $(\lambda, e)$ is the eigenpair corresponding to the dominant eigenvalue $\lambda$ of the adjacency matrix $Ad$ of the network under consideration. $ad_{ij}$ is the $ij$th element of $Ad$. $e_i$ and $e_j$ are the $i$th and $j$th components of the eigenvector $e$.

For the figure in (a), the adjacency matrix $Ad$ is

$$Ad = \begin{bmatrix} 0 & 1 & 0 & 0 \\ 1 & 0 & 1 & 1 \\ 0 & 1 & 0 & 0 \\ 0 & 1 & 0 & 0 \end{bmatrix}$$

The eigenvalues of the matrix are $\sqrt{3}$, 0, 0, and $-\sqrt{3}$. So, the dominant eigenvalue is $\sqrt{3}$ and the corresponding eigenvector is $(1, \sqrt{3}, 1, 1)^t$. So, node $B$ is more central with a centrality value that is $\sqrt{3} = 1.732$ times that of the remaining three nodes.

The adjacency matrix corresponding to the graph in figure (b) is

$$Ad = \begin{bmatrix} 0 & 1 & 0 & 0 \\ 1 & 0 & 1 & 0 \\ 0 & 1 & 0 & 1 \\ 0 & 0 & 1 & 0 \end{bmatrix}$$

The dominant eigenvalue of this matrix is $\frac{1+\sqrt{5}}{2}$ The corresponding eigenvector is $(1, \frac{1+\sqrt{5}}{2}, \frac{1+\sqrt{5}}{2}, 1)^t$. So, both the nodes $B$ and $C$ have a centrality value that is $\frac{1+\sqrt{5}}{2}$ times that of the centrality of the other two nodes.

In the case of figure (c), the adjacency matrix is

$$Ad = \begin{bmatrix} 0 & 1 & 0 & 1 \\ 1 & 0 & 1 & 0 \\ 0 & 1 & 0 & 1 \\ 1 & 0 & 1 & 0 \end{bmatrix}$$

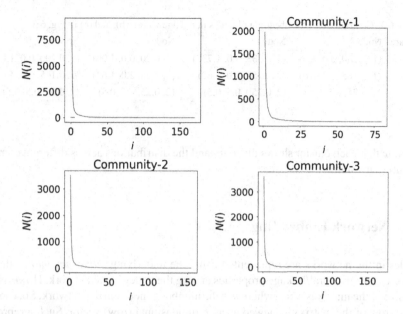

**Fig. 6.4** Node frequency versus Degree for Pubmed data set available at https://linqs.soe.ucsc.edu/data. This data set has 19717 nodes and 3 communities. It is evident that power-law degree distribution is present in the original network(top-left image) as well as in all the three communities

The dominant eigenvalue is 2 and the corresponding eigenvector is $(\frac{1}{2}, \frac{1}{2}, \frac{1}{2}, \frac{1}{2})^t$ which indicates that the centrality of all the four nodes is the same and is 0.5.

## 6.8  Community Detection

There are several applications where centrality of a node can be exploited. Because central nodes are also influential, product marketing teams try to target the central nodes in the network to improve the product promotion.

Typically, it is assumed that a central node influences other nodes in its community and the community is formed around the central node. So, diversity among these central nodes is needed to achieve a good set of communities. One of the important issues in community formation is to generate communities from a network such that the entire network $G = \{V, E\}$ satisfies power-law degree distribution and each community $G_i = \{V_i, E_i\}$ also satisfies power-law degree distribution.

This means that network is clustered into $K$ communities, where clustering a power-law degree distributed data into communities that are in turn power-law degree distributed. This is evident from the publicly available benchmark communities data sets.

**Table 6.3** Centrality-based embedding of nodes in the figures (a), (b), and (c) in Fig. 6.3

| Figure | Node A | Node B | Node C | Node D |
|---|---|---|---|---|
| (a) | $(1, 0.200, 0.0, 1.0)$ | $(3, 0.333, 3.0, 1.732)$ | $(1, 0.20, 0.0, 1.00)$ | $(1, 0.200, 0.0, 1.0)$ |
| (b) | $(1, 0.165, 0.0, 1.0)$ | $(2, 0.250, 2.0, 1.62)$ | $(2, 0.25, 2.0, 1.62)$ | $(1, 0.165, 0.0, 1.0)$ |
| (c) | $(2, 0.25, 0.5, 0.5)$ | $(2, 0.250, 0.5, 0.5)$ | $(2, 0.25, 0.5, 0.5)$ | $(2, 0.25, 0.5, 0.5)$ |

Note that each cluster shows diversity and the distributions across different communities are similar.

## 6.9  Network Embedding

Nodes in the networks are represented using the underlying *adjacency matrix* that captures the structural/linkage properties among the nodes in the network. However, the size of the matrix is $n \times n$ where $n$ is the number of nodes in the network. So, each node (row of the matrix) is viewed as an $n$-dimensional (row) vector. Such a representation can be very high-dimensional and so can be unwieldy to exploit in various network mining/analysis tasks including community detection, link prediction, and network visualization.

This prompted researchers to explore embedding nodes in a lower dimensional space. We will explore some of them next.

### 6.9.1  Node Embeddings Based on Centrality

Here, we can represent any node in any network as a 4-dimensional vector where the four components are:

1. The first component is the *degree centrality* value.
2. The second component is the *closeness centrality* value.
3. The third component is the *betweenness centrality* value.
4. The fourth component is the *eigenvector centrality* value.

We represent the nodes in each of (a), (b), and (c) in Fig. 6.3 in Table 6.3.

### 6.9.2  Linear Embedding of Nodes Using PCA

It is possible to reduce the dimensionality of the nodes using the conventional techniques like the principal components (*PCs*). For example, consider the *Ad* matrix corresponding to the graph in Fig. 6.3a. The sample covariance matrix is

**Table 6.4** *PC* Based embedding of nodes in the (a) Fig. 6.3a

| Node | 2-dimensional Representation |
|------|------------------------------|
| A | $(-\frac{1}{2}, \frac{1}{2})^t$ |
| B | $(\frac{3}{2}, \frac{1}{2})^t$ |
| C | $(-\frac{1}{2}, \frac{1}{2})^t$ |
| D | $(-\frac{1}{2}, \frac{1}{2})^t$ |

$$
\begin{bmatrix}
\frac{3}{16} & -\frac{3}{16} & \frac{3}{16} & \frac{3}{16} \\
-\frac{3}{16} & \frac{3}{16} & -\frac{3}{16} & -\frac{3}{16} \\
\frac{3}{16} & -\frac{3}{16} & \frac{3}{16} & \frac{3}{16} \\
\frac{3}{16} & -\frac{3}{16} & \frac{3}{16} & \frac{3}{16}
\end{bmatrix}.
$$

The eigenvalues of the matrix are $\frac{3}{4}, 0, 0, 0$. The two leading orthonormal eigenvectors are $(0.5, -0.5, 0.5, 0.5)^t$ and $(0.5, 0.5, -0.5, 0.5)^t$. By projecting the four 4-dimensional points (rows) in the *Ad* matrix given by

$$
Ad = \begin{bmatrix}
0 & 1 & 0 & 0 \\
1 & 0 & 1 & 1 \\
0 & 1 & 0 & 0 \\
0 & 1 & 0 & 0
\end{bmatrix},
$$

we get the 2-dimensional vectors as shown in Table 6.4.

In a similar manner, it is possible to represent the points in graphs in (b) and (c) also using their respective *PCs*. It is possible to use matrix factorization based schemes discussed in the earlier chapters to embed each of the nodes in the graph in some reduced $K$-dimensional space. However, they may be computationally expensive.

### 6.9.3 Random Walk-Based Models for Node Embedding

One of the most influential contributions to the natural language analysis models is based on representing each word in a sentence as a vector. Specifically, *Word2Vec* that was originally proposed and used in the context of natural language sentences was influential in generating node embeddings in networks. The node embedding problem was solved akin to the word embedding by conducting random walks on networks and viewing these random walks as sentences in a language made up of nodes and edges. we examine these models next.

- *Word2Vec*: In this model for each word occurring in a document, words occurring in a fixed size window around the word under consideration are selected to identify *the context* in terms of possible words that can pair up with the word. Let $w$ be the word considered and $w_1, w_2, \ldots, w_m$ be the words occurring in the context of the window around $w$. Then $(w, w_1), (w, w_2), \ldots, (w, w_m)$ are the pairs with respect to $w$. This process is repeated for all the words in all the documents in the collection to identify training pairs that capture the co-occurrence of words in the document collection.

These pairs are used to train a one hidden layer neural network. Here, the words cannot be directly input. Instead, each word is represented as a *one-hot vector* based on its index in the vocabulary $V$. If $|V| = l$, then there are $l$ words in the vocabulary. When these words are ordered in the lexicographic order, then each word in $V$ will occupy a position/index value between 1 and $l$ based on its lexicographic order.

So, a word $w$ is a binary vector of $l - 1$ zeros, and one 1; this 1 is located in the position corresponding to the index of word $w$ in $V$. So, a pair of the form $(w, w_i)$ is used to train the network by inputting the word $w$ as a one-hot vector and expecting the output to be the one-hot vector corresponding to $w_i$ as shown in Fig. 6.5. Training of the one hidden layer neural net is carried out using the pairs of words of the form $(w, w_i)$ based on the context, in terms of the window around word $w$ for every word in the document collection.

If $K$ is the number of nodes in the hidden layer, then at the end of training, each input node will be associated with a $K$-dimensional vector. So, effectively each word has an input node position associated based on the one-hot vector representation and as each input node has a $K$-dimensional representation, each of the $l$ words will end up having a $K$-dimensional representation. This is the Word2Vec representation where each word is represented as a $K$-dimensional vector. Some additional heuristics are employed to reduce the computational burden of the model; these include the following:

- using popular phrases in addition to the words in the vocabulary.
- Undersampling frequent pairs to reduce the number of training pairs.
- Negative sampling where in the output one-hot vector only a small fraction of randomly selected 0s are viewed as essential in training rather than insisting all $l - 1$ zeroes.

- *DeepWalk*: It captures the latent representations of nodes in the network by exploiting the structural information abstracted by the adjacency matrix.

  With each node, some $r$ random walks each of length $t$ are generated. A simple strategy for the random walk is to choose uniformly any one of the neighbors from the current node. Each such random walk is viewed as a sentence in some artificial language.

  Each node is initially assigned some embedding. Now, the random walks are used to update the embedding of the node under consideration by maximizing the probability of its neighbors being present in the random walk given the representation of the node. In order to make the computation feasible and efficient, the problem of probability estimation is viewed as estimating the probability of a path in a tree from the root to a leaf; the tree is rooted at the representation of the node under consideration and all the nodes are viewed as leaves of the tree. Some of the important features of DeepWalk are as follows:

**Fig. 6.5** Neural Network for Word2Vec

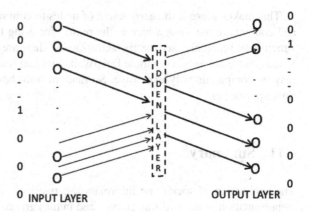

INPUT LAYER                    OUTPUT LAYER

- It is possible to perform random walks and the resulting processing in parallel. It is experimentally observed that parallel scheme for node embedding does not affect the performance of the resulting embeddings.
- Nodes sharing similar neighborhoods will acquire similar representations.
- The same embedding is found to be useful in a variety of machine learning tasks on the network like community detection, classification of nodes, and network visualization.

There are several other node embedding schemes based on first- and second-order structural neighborhood of nodes and based on breadth-first and depth-first search strategies.

## 6.10   Combining Structure and Content

We have seen so far the role of embedding nodes in a network based on structural information in the network. There are several applications where there is attribute information associated with each node in the form of content. The structural information is captured by the adjacency matrix $Ad$ which we have examined earlier in this chapter.

If $At$ is the content/attribute matrix of the network, then $At$ is of size $n \times l$ where $n$ is the number of nodes and $l$ is the number of words in the vocabulary. A solution based on matrix factorization is given by

$$Ad_{n \times n} = B_{n \times K} C_{K \times n},$$

and

$$At_{n \times l} = B_{n \times K} E_{K \times l}.$$

This makes sense if the assignment of nodes to communities based on structure and content are the same which is the reason for using the same $B$ matrix in both structure and content based matrix factorization. There are other matrix factorization-based combinations that combine DeepWalk with text features. Matrix factorization may be computationally expensive. So random walk based methods might be the right approaches.

## 6.11  Summary

The application of social and information networks is considered in this chapter. Representation of networks as graphs, and in turn graphs as adjacency matrices are considered. The important notion of centrality in networks and several variants of centrality in networks was considered. Also, the notion of diversity in the benchmark communities in real-world networks was explored. Representation of networks both in terms of structure and content was explored.

## Bibliography

1. Virinchi S, Mitra P (2016) Link prediction in social networks: role of power law distribution, Springer Briefs in Computer Science, Springer
2. Leskovec J, Rajaraman A, Ullman J (2014) Mining of massive datasets. Cambridge University Press
3. Hasan MA, Chaoji V, Salem S, Zaki M (2006) Link prediction using supervised learning, In: Proceedings of SDM 06 workshop on counter terrorism and security, 20–22 April, 2006, Bethesda, Maryland, USA
4. Liben-Nowell D, Kleinberg JM (2003) The link prediction problem for social networks, In: Proceedings of CIKM, 03-08 November 2003, New Orleans, LA, USA, pp 556–559
5. Sambaran B, Ramasuri N, Murty MN (2018) A Generic Axiomatic characterization for measuring influence in social networks. ICPR 2018:2606–2611
6. McCormick C (2016) Word2Vec Tutorial—The Skip-Gram Model http://mccormickml.com/2016/04/19/word2vec-tutorial-the-skip-gram-model/ (2016)
7. Perozzi B, Al-Rfou R, Skiena S (2014) DeepWalk: online learning of social representations, KDD, August 2014, New York
8. Samabaran B, Lokesh N, Murty MN (2019) Outlier aware network embedding for attributed networks, AAAI 2019, Honolulu
9. Yang C, Liu Z, Zhao D, Sun M, Chang EY (2015) Network representation learning with rich text information, IJCAI 2015, Buenos Aires

# Chapter 7
# Conclusion

In this book, we have examined the role of centrality and diversity in search. Specifically, we consider the roles of centrality and diversity in

1. Search, representation, classification and clustering, ranking, and regression in an introductory manner. The roles of bias and variance in regression is considered in detail; the correspondence between centrality and diversity against variance and bias is also examined.
2. Search in detail is considered in Chap. 2. Variations like exact and inexact search are considered. Searching for proper representation, proximity and distance functions, clustering and classification, information retrieval, and AI problem-solving are considered. The roles of centrality and diversity in search-based applications is summarized.
3. Representation is considered in detail in Chap. 3. Its importance in AI problem representation, document representation, clusters, classes, and classifiers is examined. The roles of centrality and diversity in a variety of representation-based tasks is summarized.
4. Clustering and classification is considered in Chap. 4. Specifically, the role of optimization and regularization and their relation to diversity and centrality in representation, clustering and classification is summarized.
5. Ranking is considered in Chap. 5. Ranking based on similarity and density is considered. The roles of centrality and diversity in ranking, summarization, and recommendations are examined.
6. Social and information networks in Chap. 6. A detailed discussion on representation of networks, link prediction, centrality, community detection, and network embedding are considered in detail. The roles of diversity in these tasks is considered. The role of centrality, linear transforms, and random walk-based models in node embedding in networks is also considered.

M. N. Murty and A. Biswas, *Centrality and Diversity in Search*,
SpringerBriefs in Intelligent Systems, https://doi.org/10.1007/978-3-030-24713-3_7

# Glossary

| | |
|---|---|
| $\Sigma$ | Covariance Matrix |
| $C_+$ | Positive class |
| $C_-$ | Negative class |
| $W$ | Weight vector |
| $b$ | Threshold weight |
| $\alpha$ | weight of support vector |
| $\mathcal{L}$ | Lagrangian |
| $X_i$ | $i$th pattern |
| $y_i$ | Class label of the $i$th pattern |
| $G$ | Graph representing a network |
| $V$ | Set of vertices or nodes in the graph |
| $E$ | Set of edges in a graph |
| $\pi(\mathcal{X})$ | Hard partition of the the data set $\mathcal{X}$. |

# Index